YOU CAN SUPPORT
CRUELTY-FREE PRODUCTS...
IF YOU KNOW WHICH PRODUCTS TO BUY

Acne Products • Aftershaves and Colognes •
Air Fresheners • Antiperspirants • Automobile
Products • Batteries • Bleach • Carpet and Rug
Products • Cold and Flu Products • Dental
Floss • Drain Openers • Eye Drops • Face and
Skin Care Products • Glass Cleaners • Insect
Repellents • Kitty Litter • Lighters • Paint and
Enamel • Razors • Shampoos • Shoe Polish •
Suntan Products • Toys • Vitamins

All these and much more are listed in...

A SHOPPER'S GUIDE TO
CRUELTY-FREE PRODUCTS

A
SHOPPER'S GUIDE TO CRUELTY-FREE PRODUCTS

BY
LORI COOK

PRINTED ON RECYCLED PAPER

BANTAM BOOKS
NEW YORK · TORONTO · LONDON · SYDNEY · AUCKLAND

A SHOPPER'S GUIDE TO CRUELTY-FREE PRODUCTS
A Bantam Book / June 1991

ISBN 0-553-29208-0

Published simultaneously in the United States and Canada

Bantam Books are published by Bantam Books, a division of Bantam Doubleday
Dell Publishing Group, Inc. Its trademark consisting of the words "Bantam Books"
and the portrayal of a rooster, is Registered in U.S. Patent and Trademark Office
and in other countries, Marca Registrada. Bantam Books, 666 Fifth Avenue, New
York, New York 10103.

PRINTED IN THE UNITED STATES OF AMERICA

OPM   0 9 8 7 6 5 4 3 2 1

For the innocent victims of product testing—
may this help end their misery

Whenever people say, "We mustn't be sentimen-tal," you can take it they are about to do something cruel. And if they add, "We must be realistic," they mean they are going to make money out of it.

Brigid Brophy

The question is not
Can they reason? nor
Can they talk? but
Can they suffer?

Jeremy Bentham

CONTENTS

C

D

H

I

J

K

L

M

N

O

P

FOREWORD

by Priscilla Feral
President, Friends of Animals

The book you are about to read is the product of a quiet revolution that has been waged in the grocery stores and shopping malls of America over the past five years. It might be called the revolt of the "green" or "cruelty-free" consumer. It has irrevocably changed the consciousness of shoppers across America. And it has ushered in the decade of the "cruelty-free" product.

Why has this revolution been so effective and so far-reaching? The answer lies partly in the growing awareness of the gross exploitation of animals in research laboratories across the country. The American consumer is finally saying NO to such barbaric and cruel tests as those designed to put toxic products into the eyes of conscious rabbits or to force-feed oven cleaner to innocent beagles. Such experiments are horribly cruel to animals. Every year nearly 15 million animals suffer and die due to experiments that purportedly determine the "safety" of household and cosmetic products. Animals experience convulsions, bleeding from the nose and mouth, constricted breathing, emaciation, horrible pain, and lingering deaths. American consumers are leading the world's shoppers in saying that such pitiful animal suffering is too great a price to pay for a new color of nail polish or a new deodorant.

The consumer revolt, though, has not only been

for animal and human safety. It has also been for the good of the earth. American shoppers realize that products with chemical ingredients are contributing to the pollution of the earth at an alarming rate. Sunset over the Pacific is not a pretty sight when a five-inch layer of detergent scum continually washes up at the viewer's feet! Indeed, the fight for natural, biodegradable ingredients will do much to restore our living ecosystem to its proper balance.

Furthermore, many products contain animal ingredients. Protein-enriched shampoos, for example, are made with animal by-products, glandular extracts, and sometimes fetal fluids. The tallow that is in most commonly used soaps is derived from chemical- and hormone-fed cows, pigs, and chickens and also frequently from cats and dogs who have been euthanized at pounds and shelters. American shoppers are opting for healthy, life-giving ingredients derived from natural plant products instead of ingredients resulting from animal mutilation and death. The massive consumer demand for products that are ecologically sound is moving many companies to use alternatives to animals. And consumer outcry will continue until the last of the recalcitrant manufacturers "sees the light."

Consumers can and are preventing the suffering of millions of animals in product-testing laboratories around the country and around the world. By not using products that are animal-tested, they are also ensuring their own health and the present and future health of this planet. The quiet revolution that is ending product testing on animals has already had many victories. Companies such as Benetton, Avon, Noxell, and others have tumbled to consumer pressure and agreed to stop animal testing. Others have issued moratoriums on the use of animals — and shoppers are using their dollars to ensure that these termporary bans become permanent ones.

So use this book in good health and know that you are also contributing to the health and well-being of animals and, in fact, of the entire planet. Armed with the information you are about to read, you are ready to become a "cruelty-free" shopper. And you are about to take a step into the "green" and living future of the earth.

ACKNOWLEDGMENTS

There are many people I want to thank for making this book a reality. First of all, I want to thank my agent, Nancy Love; my editor, Leslie Meredith; and her assistant editor, Gina Velasquez, for having the willingness to take a chance on an unknown writer with an idea. Friends of Animals and Priscilla Feral also deserve thanks for sponsoring this book and writing the foreword. I hope I don't let any of these wonderful people down.

Next in line for thanks are all the companies who participated and managed to get the questionnaires back on time despite a very short deadline. I really couldn't have done it without them.

Three people who really deserve thanks (and they know it) are the employees of the Tampa CompuAdd Computer Store—Jim, Gary, and Kelli. Despite daily frantic calls to them, they managed to keep their sense of humor and get me through this. This also applies to the guy in the New Mexico store who saved me when the stores in my time zone were closed.

And finally I want to thank my family, for they are the ones who first taught me to love all living things.

A SHOPPER'S GUIDE TO CRUELTY-FREE PRODUCTS

Man's inhumanity to man is only surpassed by his cruelty to animals.

INTRODUCTION

Look around your home. Look at the products in your bathroom cabinet, under your kitchen sink, or in your garage. Has cruelty been on your shopping list? Is animal pain, suffering, and death associated with the products you buy? Chances are your shampoo poisoned a cat and your bleach was used to blind a rabbit. Do you plan on injecting dish soap into your bloodstream or eating your deodorant? Do you need to know the consequences of these kinds of actions before you use a product? Some manufacturers think so, and they kill millions of animals every year in an attempt to answer these ridiculous questions. They call it product testing; others call it cruelty.

Animal product testing is the practice of using live, conscious animals to determine the amount of damage various consumer products can have on living tissues. This is accomplished by poisoning the animals and is in no way related to medical research or drug trials. This practice is one that manufacturing firms elect to perform on their consumer products before, during, and after the sale of a product. These tests are not required by any law for most consumer products. Furthermore, they have been shown repeatedly to be inaccurate, unreliable, useless, redundant, and incredibly expensive. But most of all, these tests are extremely cruel. Despite alternatives that are less expensive, more accurate, and do not require the use or death of animals, some companies refuse to stop these tests. Apparently, it is up to the consumer to put an end to this barbaric practice.

The various products available to consumers are regulated by several government agencies. The Food and Drug Administration (FDA) controls cosmetics, while most remaining consumer products are covered by the Consumer Product Safety Commission (CPSC). Insecticides, fungicides, rodenticides, and some toxic products such as solvents are regulated by the Environmental Protection Agency (EPA).

The FDA does not require animal tests for any products that are not foods or drugs (such as, health and beauty aids or cosmetics). The EPA requires the manufacturers of insecticides, fungicides, and rodenticides to register their products with them. When doing this, the manufacturers must include information on the health aspects (to humans) of their various products. They are not required by law to use animals to determine this data. If the EPA feels the data is inadequate or suspect or if an existing product is being reregistered for a new use, the EPA may require additional testing. This additional testing may require the use of animals. The CPSC does not require animal tests for consumer products unless covered by federal acts such as the Federal Hazardous Substances Act and the Federal Insecticide, Fungicide, and Rodenticide Act. No agency requires routine animal tests for health and beauty aids or household products.

The Requirements of Laws and Regulations Enforced by the U.S. Food and Drug Administration reads as follows: "Although the Federal Food, Drug, and Cosmetic Act does not require that cosmetic manufacturers or marketers test their products for safety, the FDA strongly urges cosmetic manufacturers to conduct whatever toxicological or other tests that are appropriate to substantiate the safety of the cosmetics. If the safety of a cosmetic is not adequately substantiated, the product may be considered misbranded and may be subject to regulatory action unless the label bears the following statement: *Warning—The safety of this product has not been determined (21 CFR740.10).*"[1]

In essence, this says there is no law that requires any testing in general or animal testing specifically. The individual companies are expected to decide what testing, if any, they will do. The companies that perform animal tests, or contract with others to perform them, do so of their own free will and not because of government regulations. They have chosen to use animals despite the pain, death, incredible cost, and availability of other alternatives.

Yet more than 500 companies have managed to develop, test, and manufacture products without killing animals or endangering the public's safety. Many are listed in this book.

Companies that insist on animal testing test anything from shampoo and drain opener to eye shadow and crayons. One manufacturer even tests its blue jeans for irritation by applying fabric patches on the shaved, exposed, raw backs of rabbits.[2] These tests are referred to by those who practice them as "safety" tests, but this is a misnomer used to get the public to accept this practice.

These tests in no way make the products safer for humans. The animals receive no medical help during or after these experiments, so no medical knowledge is gained in how to treat poisonings, eye inflammations, or toxic problems caused by accidental misuse of a product. The results of these tests are considered trade secrets and usually are not even made available to the government, much less the public or medical fields. Furthermore, if these tests were indeed being done to "ensure our safety" as the users of these tests like to say, one would assume that unsafe, toxic, or dangerous products would not make their way to the stores' shelves. Unfortunately, nothing could be further from the truth. All too often the product is marketed in spite of the results of these tests, as in the case of AETT (acetyl ethyl tetramethyl tetralin).

A popular ingredient, AETT was used in everyday products such as deodorants, soaps, and aftershaves. In 1975, during routine animal tests, a manufacturer found that AETT was extremely dangerous and toxic. It caused severe nerve damage, tissue discolorations, and death to test animals, even after a single topical application or inhalation. is was a very dangerous and unsafe ingredient, yet the manufacturer did not immediately recall and reformulate all products containing AETT. It and others in the industry who knew of AETT's toxicity continued to "study" the "problem," killing more animals as they did, for three more *years* while continuing to sell products containing this ingredient. In 1977, three years after the first animals died of this toxic substance, this manufacturer and others finally decided to stop using AETT but still felt no need to recall products that contained this ingredient.[3]

This is not the only instance, ingredient, or company that continued to sell products after testing proved them unsafe. The headlines are filled with recalls and with lawsuits claiming that companies ignored the results of their testing. In 1988, 35,000 people were treated in emergency rooms across this nation for cosmetic-related injuries.[4] A look through Ruth Winter's *Consumer's Dictionary of Cosmetic Ingredients* shows many known carcinogens and irritants that are still in use. Toni Stabile's book *Everything You Want to Know About Cosmetics* is filled with stories of dangerous products that made it to the marketplace despite animal tests. When the FDA tries to curtail or prohibit the use of a carcinogenic ingredient, the companies that use the ingredient in their products usually try to prevent the banning.[5] Where is their concern for the public's safety then?

No one wants to use an unsafe product. Product tests are a good idea, but only if the test is an accurate indicator of the product's safety. Animal tests have proved repeatedly they are not.

It should come as no shock to anyone, but animals and people are different. Animals also differ among themselves. Any cat owner can tell you that. Some products that are safe for use on a dog will kill a cat. Likewise, products that have tested safe on animals have been known to kill people. While product testing is not related to drug testing, the stories of penicillin and thalidomide illustrate this point very well. Luckily, penicillin was developed and tested without using animals—luckily, because penicillin is extremely lethal to guinea pigs.[6] If it had been animal tested and the manufacturer had heeded the results of those tests, its immense gift to mankind might never have been recognized. The drug thalidomide shows the other side of the story. Thalidomide, which many women took for nervousness and nausea during pregnancy resulting in tragic birth defects, tested safe on animals.[7] Clearly, animal tests are not a good indicator of the safety of a product when used by people.

Not only are these animal tests inaccurate indicators of a product's safety—that will probably be ignored by the companies anyway—and not required by law, they also generate useless information. For some substances, the lethal amount varies widely among species. The amount needed to kill a dog could be hundreds of times higher or lower than the figure that would kill a person. Knowing these figures will have no bearing at all on people. Doctors faced with an accidental poisoning need to know how to help the victim. Knowing how much of the product it took to kill a cat will be of no use.

Dr. Christopher Smith, FACEP, medical director of the department of emergency medicine at Dominguez Medical Center in Long Beach, California, has said, "As a board certified emergency room physician with over 17 years of experience in the treatment of accidental poisonings and toxic exposures, I have never used animal tests to manage accidental poisonings; to do so would be irrational."[8] Dr. Herbert Gundersheirmer claims, "As a physician with 45 years experience in internal medicine, I can find no justification for these cruel animal tests. The results are not transferrable from species to species, and therefore cannot guarantee product safety."[9]

If animal tests are not required by law and they don't ensure our safety and if they are not reliable indicators of human toxicity and the results are just useless figures, why are they still being performed? Usually, these tests are done to appease the legal departments of the big manufacturing firms. In case of a legal dispute over the use or misuse of a product, the manufacturer likes to have some document that will prove its position in the dispute. This is clearly illustrated by the following story from an article in *PETA News*.[10]

A woman's eye was allegedly damaged after using a L'Oreal product. In response to this, L'Oreal tried to prove that the damage to the consumer's eye was actually caused by water when she flushed her eye, not by their product, even though flushing one's eye with water is an accepted first aid practice. In an attempt to prove this, they paid a research firm to hold a live rabbit's eye open under a stream of running water for two solid hours, far longer than the woman flushed her own eye. Needless to say, the poor rabbit's eye did indeed become damaged, but this does not prove that either the water or the L'Oreal product damaged the consumer's eye. To me, it does prove L'Oreal's disrespect not only for animal life but for their customers as well.

The Animals

Every three seconds another lab animal dies. According to the Institute for the Study of Animal Problems, 100 million animals a year are used for research and testing in this country. Laboratories use 50 million mice, 20 million rats, and more than 30 million other animals, including dogs, cats, and horses. There is no way to know exactly how many of these are used in worthless cosmetic tests, but the number is estimated at one-fifth, or 20 million, animals.[11]

Product tests are performed on dogs and cats, rats and mice, rabbits and guinea pigs, even primates and pigs. Horses, cows, sheep, even hyenas have been used. Virtually any living animal is acceptable to the laboratories, and the source for these animals are varied and many. Most arrive at laboratories via a middleman called a dealer or buncher.

Dealers work with small domestic animals or large exotic ones. Some specialize; others do not. Some work within the law, some don't. Some animal dealers breed animals specifically for laboratories, those that do not breed animals buy them from "random sources" such as zoos, animal shelters, and individuals who sell stolen pets.[12]

Surplus zoo animals can end up in testing laboratories. Many zoos deliberately breed primates for which they have no space simply to keep the nursery or petting zoo filled with cute animals. Because primates are very intelligent and emotional animals, deliberate overbreeding causes much emotional trauma to both mother and child when they are separated. When the animals are no longer cute and cuddly, they are sold to dealers who may sell them to labs.

One of the most disturbing ways dealers and testing facilities acquire animals is through the use of pound seizure laws. In some areas surplus pet animals in pounds and shelters are sold or given to dealers or directly to the laboratories and research facilities. In some parts of the country this is mandated by law, while in others the laws make it illegal. What many people wrongfully assume is that an unwanted pet, surrendered at the local animal shelter, will either find a new home or painlessly be put to sleep. Unbeknownst to them, their pet may be dying a slow death from drain opener poisoning inflicted at the hands of a manufacturer whose products sit under their kitchen sink.

Even more upsetting than pound seizures is the practice of pet theft. Dealers patrol neighborhoods and easily take loose or stray animals. Some pay children or others to bring them stolen pets. Any pet is fair game. Loose pets are the best; friendly ones make the job easier. It makes no difference if your animal is wearing a collar or tied in your yard or behind a fence. If you are not supervising your pet, collars will be removed, chains untied, and gates opened. More than two and a half million pets are stolen every year in this country. Most end up in research labs.[13]

Another very easy way of acquiring animals for resale to laboratories is the "Free to Good Home" ads. Dealers pick up many animals in one day simply by using these ads, all the while explaining what a great home they will give a dog or litter of kittens. They might hit every ad in the paper in one day and return for more the next day, telling of a neighbor or relative who saw the first animal and now "wants one just like it."[14]

Once they are with the dealers, an animal's nightmare has just begun. Some will be kept in deplorable homemade cages, boxes, or bags until sold. Some dealers won't bother cleaning their cages or feeding them since this increases overhead. Others will be transported thousands of miles in hot, cramped trucks. Depending on their species and the facility at which they end up, things might get better, temporarily, or they might be killed simply because they do not meet the right specifications for the test (such as, size, weight, age).

The United States Department of Agriculture's (USDA) Animal and Plant Health Inspection Service (APHIS) is supposed to investigate complaints of animal cruelty against these dealers, but they do very little. In one instance, a dealer committed 25 alleged violations in 32 months yet never received anything more than a warning to correct the problem from the USDA.[15] In Missouri, which has more licensed dog dealers than any other state, auctions of animals are a common occurrence. Dealers looking to buy animals regularly attend these events, which often include stolen pets. Yet the APHIS animal care regional supervisor admits never even attending one of these auctions.[16]

In 1966, Congress passed the Laboratory Animal Welfare Act to protect pets from being stolen for research as well as to protect those animals already in labs. Congress gave the United States Department of Agriculture (USDA) power to enforce this act. Eventually the act was renamed the Animal Welfare Act (AWA). The original intentions of the act were noble, but the AWA, as written today, is full of loopholes and wording that does little to protect animals either in or out of testing laboratories.

The Animal Welfare Act as written by Congress originally covered dogs, cats, guinea pigs, hamsters, rabbits, and all other warm-blooded animals used in biomedical research and testing.[17] Yet the USDA decided to exclude birds, mice, and rats from protection under the act. Birds, mice, and rats make up 84% of animals used in these places.[18]

The fate of the remaining 16% is not much better. The research community has requested that the USDA exclude gerbils from AWA protection as well. To this request the USDA responded they did "not have the authority to remove these animals from coverage of the regulations."[19] If they do not have the authority to remove gerbils from the list, why they had the authority to remove birds, rats, and mice is not clear.

Other aspects of the Animal Welfare Act are just as ridiculous. The AWA specifies that animals are to be treated humanely before and after an experiment but does not protect the animals during it. Dogs must be given exercise, but no concern for their psychological well-being has to be considered. On the other hand, primates' psychological well-being has to be considered, but they don't have to be exercised. Cages must meet minimum size standards as set by the AWA, but laboratories can continue using existing under-sized cages until they wear out. The lifespan of these cages is approximately 25 years, so no changes are expected in this decade.

The loopholes and inadequacies of the Animal Welfare Act goes on and on. It in no way protects animals from cruelty, especially if the animals are already in a testing facility. But it does seem to protect the testers themselves. Cora Rudek of Oklahoma received 30 days in jail and a $250 fine for poisoning a cat with Drano.[20] Drackett, the manufacturer of the product, and any other manufacturer, can do this without facing cruelty charges by saying it's being done as a product test.

When used for the various tests, very few, if any, of these animals will receive painkillers or anesthesia. Most of the companies that choose to do these tests also choose not to alleviate the suffering and immense pain the animals feel as they are poisoned. The companies claim that medications might interfere with the test but do not mention that these tests are not an accurate indicator of anything other than the "size and dose required to commit suicide."[21]

Using anesthetics, painkillers, or even antihistamines would provide some relief to these animals. While some drugs may interfere with tearing or blinking, many others do not.[22] These other medications could and should be used by those companies who refuse to stop using animals in their routine tests. It is, literally, the least they could do.

In some instances, if the animal survives one test, he may be used again in another type of product test. Rabbits who have one eye destroyed in a Draize test may be used again in a topical irritation study. Eventually, all the animals used will die or be killed.

One way to protect pets from this type of fate is to tattoo them. Collars can get lost or removed, but tattoos cannot. Most testing and research facilities attempt to return lost and stolen pets to their owners. The only permanent way to identify your cat or dog is to have it tattooed on the inner thigh (tattoos in ears are removed by cutting the ear off). This is a painless, ten-minute procedure that does not require anesthesia. Once registered with a pet registry firm, your pets will always have identification that will enable them to be returned to you. In early 1988, an employee at a medical research facility noticed a tattoo on a dog scheduled for experimentation. Tracing the tattoo number he found the dog's owners and reunited them. The dog had been stolen from their backyard eight months earlier.[23]

Some state legislators are attempting to pass laws that would make it illegal for pounds and shelters to sell pets for experimentation without the owners' consent. Others are trying to get bills passed that would make it illegal for laboratories to use animals if they have tags, licenses, or tattoos. As tags and licenses can be easily removed, tattoos would provide the only permanent protection for the pets and their owners. Unfortunately, mice, rabbits, and zoo animals can't receive the same protection.

The Tests

Products are tested on animals in any number of ways. The test substance is forcibly ingested, inhaled, or injected, applied topically to the skin, or dropped in the eyes and ears. It might be put it in the vagina or the rectum, mixed with food, added to water, or force-fed directly into the stomach with a tube. For topical toxicity, testers shave the animals, scrape the skin, then apply the test substance. Chronic studies use small amounts over long periods of time, while acute tests use massive quantities all at once. None of these tests uses a product in the function for which it was developed. They do not, for instance, test shampoos by washing the animal's fur.

Animal testing is very expensive. According to the General Accounting Office, the average cost to test on animals is half a million dollars per product. The average cost for nonanimal testing is roughly a tenth of that, at $50,000 per product.[24] Consumers, of course, ultimately pay for this incredible expense.

LD/50

One of the oldest and best known tests is the LD/ 50. This stands for the "lethal dose" of a given substance that will definitely kill 50% of the group of animals on which it's tested. This is a test for toxicity—to determine how poisonous a product is. This test can also be performed to find the lethal dose to more or fewer animals, referred to as LD/80 or LD/40 respectively. If the substance to be tested is a gas, the test is referred to as the lethal concentration (LC). The test group can range from 6 animals up to hundreds.

The LD/50 test is done on rats, rabbits, primates, pigs, exotic animals found usually in zoos, and common ones like a family's pet dog or cat. Usually no painkillers or anesthesia are given before, during, or after the test.

The products to be used for this test can be administered in any number of ways but usually are forced directly into the throat or stomach, injected into the abdominal cavity, or applied topically. They also can be injected into a vein, applied to the eyes, rectum, or vagina, or forcibly inhaled.

The animals allotted for testing a specific product are divided into groups. The administrators give each group increasing doses or quanities until the designated percentage of animals die in the designated period, usually a few days to a couple of weeks. If the product is a highly toxic one, such as drain opener, weedkiller, or bleach, the quantity and time needed to kill the group of animals will be small. If the product is not toxic, some testers will continue to increase the quantity of it, in an attempt to get the required kill figures, until the animal literally explodes from the sheer volume.[25]

Once the animals have been dosed with the

product or ingredient, they are put back in a cage and monitored or watched at various intervals for the selected period of time. The animal will receive no help to survive or die. Depending on the substance and the amount given, the animals may convulse or hemorrhage from the mouth, nose, ears, eyes, vagina, or anus. They can lose consciousness, become paralyzed, vomit, and get diarrhea. They can cry, shriek, howl, and moan. They can die quickly, or they can die a slow, lingering death. If they manage to live through the test, they probably will be killed by the test administrators anyway for examination of the internal organs. Some survivors will be used in other tests until they eventually succumb to a lethal dose.

This test is very crude and incredibly inaccurate. The results vary enormously depending not only on the species being poisoned but even individually within the same species. The results can be affected by age, sex, stress, and general physical condition of the animal, just as human reactions to the same drug or stimuli differ. Even the type of bedding and the time of day can affect this test.[26] Animals are affected by circadian rhythms just as we are. These rhythms have dramatic affects on test results. At one time of day a fixed dose might only kill 10% of the test group, while the same amount given at a different hour would kill nearly 100%.[27] The method used to poison the animals also will directly affect the results. A substance may be very lethal if injected into the bloodstream but not if applied topically. It has already been shown that the results obtained on one species will have no bearing on another.

Scientists, researchers, doctors, and animal lovers have all questioned this test. J. K. Morrison, author of *Modern Trends in Toxicology*, wrote in 1968 that "the resultant information contributes little to . . . safety in man," while others state, "there is no justification for using the LD/50".[28] A consulting toxicologist to the World Health Organization called this test "a ritual mass execution of animals."[29] Even groups that advocate using animals in research such as the National Society for Medical Research, say "the routine use of the quantitative LD/50 test is not now scientifically justified . . . "[30] In 1981, another group of experts called this a "near useless" test.[31] The list of experts on both sides of the animal rights controversy who are against this type of testing is endless, yet the legal departments of some of our largest consumer products companies routinely continue to practice this test.

The Draize Eye Irritancy Test

Another common test is the Draize Eye Test. In this test, rabbits (usually) are placed in stocks with only their head sticking out of one end. The lower eyelid of one eye is pulled out to form a pocket, and the substance to be tested is dropped in. By applying product to only one eye, the tester can use the remaining eye as a control by which to judge the damage. The eye can be held open with clips so the substance cannot be blinked away. As rabbits do not tear as humans do, the irritating substance cannot be washed away naturally either. The restraint boxes in which they are encased effectively stop them from rubbing at the eye. Depending on the substance, the pain can be extreme. Rabbits have broken their backs struggling in the restraint boxes to escape the gross destruction of their eyes. No painkillers or anesthesia are used.

This test usually lasts from one to three days up to two weeks, with the rabbits locked in the box the whole time.

At certain intervals, the rabbits' eyes will be examined, and the irritation, ulceration, hemorrhage, or blinding will be noted. As with most other animal tests, the rabbits are usually killed after the tests regardless of the damage to the eyes.

Like the LD/50, this test is also unreliable. Dr. Stephen Kaufman, a noted ophthalmologist from New York University Medical Center, says, "I have no use for Draize test data because the rabbit eye differs from the human eye." He went on to say, "I know of no case in which an ophthalmologist used Draize data to assist in the care of a patient."[32] In 1971, investigators looking into this test found "extreme variation" in the way the various laboratories evaluated what they saw and concluded that these tests "should not be recommended as standard procedures." They went on to say that "these tests result in unreliable results." When researchers rechecked the above investigators' findings, they found "striking differences in the way different technicians reported what they thought they saw in rabbits."[33] It's been 20 years since then, and some manufacturers are still using this test, which is so crude that it doesn't even hold up in court. Yet that is the main reason some manufacturers use it.

In 1974, an Ohio court ruled against the FDA in favor of a manufacturer. The FDA had sued the company after a girl suffered eye damage using one of the company's products. To prove that the product was irritating, the FDA did the Draize test. The court found that results on rabbit's eyes could not be extrapolated to humans, and the manufacturer won. Sixteen years later manufacturers are still performing this "test" in case of legal disputes, despite the ruling that it is of no use in human eye damage claims.[34]

The Alternatives

Alternatives to the use of animals in product tests are available and are currently being used across this country. The approach to alternatives and the development of alternatives can be broken down into three areas referred to as the three R's—reduce, refine, and replace.

The aim of this approach is to *reduce* the amount of animals used in testing, *refine* the tests themselves, and eventually *replace* all animal tests with non-animal alternatives. While this may take a while when referring to medical or drug research, it is currently available in consumer product testing.

As we have seen, the FDA not only does not require these product tests, they do not require that these firms make their results or the safety data available afterward.[35] This results in different companies performing the same tests on the same ingredients simultaneously. This further wastes the lives of animals. If these companies refuse to eliminate animal tests entirely, they could make refinements to eliminate the need for multiple tests. Unfortunately, this will probably never happen because companies view products and test results as trade secrets, which are not to be shared despite the unnecessary and cruel consequences.

Common sense is an often overlooked alternative to animal tests. Anyone with any intelligence knows that if you get soap in your eye, it will sting. Do we really need to "prove" this repeatedly with rabbits? Also, no one who does get soap in his or her eye would leave it there for twenty-one days. Such a test on an animal contributes nothing to the public's safety. By using common sense and placing the appropriate warning on the product, animal lives and the expense of testing for the company would be saved.

Another option available to the companies that still use animals in product testing is simply to stop. No law mandates these tests. Hundreds of companies have found ways to manufacture safe products simply by using ingredients that have already been tested and successfully used for years. If these companies can do it, there is no reason the others can't as well.

Ingredients known to be safe are called "Generally Recognized As Safe" (GRAS). In 1958, Congress established a list of GRAS ingredients. Since then, thousands of tests have been done on these numerous ingredients and substances, all which can be used in product development without the need for more animal tests. The Cosmetic, Toiletry and Fragrance Association (CTFA) has a list of recognized safe ingredients as well.

For companies who choose to use ingredients that are not on the GRAS or CTFA lists, alternative tests are available and currently in use by other consumer corporations that do not require the death of animals. Most of these alternatives are less expensive and more accurate and reliable than those that use whole animals.

The agarose diffusion test is a nonanimal test used instead of the Draize test by Noxell, the parent company of Noxema and Cover Girl.[36] * Over 60 companies use the Eytex Test, another nonanimal testing alternative for the Draize, including the nation's largest cosmetic firm, Avon, and also Mary Kay Cosmetics.

Cell and organ cultures are also being used successfully instead of the Draize test. This test uses corneal tissue from humans or rabbits as well as whole organ cultures and has been found to mimic the results found when using live animals. It has produced more accurate findings than the Draize test.[37]

The Chorioallantoic Membrane (CAM) test uses fertilized chicken eggs to test for irritation. Other alternatives are computer and mathematical models.

The alternatives to animal testing are available and effective. Statistics show that costs decrease for companies using these and other alternatives.[38] The cost of a two-year chronic toxicity test in rats is over half a million dollars.[39] Multiply that by the number of products any corporation has, and you will come up with not only an astronomical figure but an incredible length of time wasted waiting for test results. All the alternatives currently available or in development are cheaper and faster.

If the tests are inaccurate, unreliable, and useless, if they are also cruel, unnecessary, and redundant, if alternatives are available that are more humane, less expensive, and more accurate, and if precedent has been set that animal tests cannot be

* While Noxell claims not to have performed any animal tests for the last two years, they refuse to commit to even a six-month moratorium on animal testing in the future. Therefore, they are not included in this book.

extrapolated to humans, why do some companies insist on performing them? No one really knows. Habits, legal departments' insistences, and the cost of revamping laboratories and retraining personnel may all be influencing those who insist on animal testing.

In the last few years many large consumer products companies have been forced to stop testing on animals by consumers and their buying power. It appears this is the only way the remaining firms will stop this barbaric practice. We can put a stop to this. We must work together. One person can make a difference. Will you?

GUIDELINES FOR THIS BOOK

One day my friend Stephanie asked me which mascara was cruelty-free. Two days later my neighbor Karen asked me the same question about deodorants. Both these questions came shortly after an irritating journey to the grocery store in which I could not find a nonanimal-tested product I needed. And so the idea for this book was formed.

While writing a book that would list companies and products that do not test on animals sounds like a simple task, the research proved that it was anything but. Though it appeared to be an easy task to divide companies that do test on animals from those that don't, I soon found out otherwise. It became apparent that it was not going to be a simple issue to decide which companies would or would not be included. Clearly, some arbitrary lines had to be drawn. While others may or may not agree with where I have chosen to draw those lines, I feel that all interested in this issue should understand my choices.

The first line to be drawn was an obvious one. No company that currently tests its products on animals would be included. All the companies listed in this book were sent a questionnaire in order to verify and clarify their own position, as well as the position of their parent and subsidiary companies, on animal testing. All those included have stated in writing that they do not test their products on animals.

Some companies, which are not included in this book, will state that they don't test their products on animals, which, technically, is true. But they have others do it for them. Therefore, the first question in the questionnaire was phrased to include outside testing.

The second area with which I was concerned is the ingredients used to make the finished product. One or all of the various components that make up each product could currently be tested on animals. It was beyond the scope of this book to investigate not only every company, but also every supplier the companies deal with, to determine which ingredients might be tested on animals. Therefore, I drew another line here. While the companies included in this book do not currently test on animals, the status of their suppliers is unknown and was not investigated. Some of the companies require that their suppliers do not use animals in testing, some do not.

Another interesting question that came up was: Should only products that have never been tested be included, or would it be acceptable to include those that were tested in the past but not currently? Many companies have never tested their products on animals, and others that used animals in the past chose to stop voluntarily. A few companies, however, have had to be dragged, kicking and screaming, into enlightenment.

I decided, however, that the real issue was not *when* or *how* these companies saw the light but rather that they *did*. Therefore, some of the products listed here have indeed been tested on animals in the past.

While no product listed is currently being tested on animals, there is no guarantee that they might not be tested in the future. It is now commonplace for mergers and acquisitions to change the status of a whole company overnight. All the companies were asked if their no testing policy would change in the next six months, and all included have said no.* This should ensure that the book at the time of printing is accurate to the best of my ability. I suggest that those interested keep their eyes and ears open to possible mergers and takeovers and research the status of the parent company.

A very touchy area that demanded attention was the terminology of those companies that have stopped animal testing. Some have stated that they have "permanently" stopped, while others have only instituted a "moratorium" on these tests. With *Webster's New Collegiate Dictionary* defining moratorium as a "period of delay" or "waiting period," it was difficult to determine whether or not to include these companies in this book.

In the end, I decided that they indeed deserve to be admitted. These companies are using the moratorium to judge the public's opinion on this issue. If their product sales increase during the moratorium, they will more than likely make the decision permanent to end all animal tests. This is the result we all want to see, so to ignore or criticize them for only making a small first step does not seem fair. It would be advisable to monitor their status from time to time but, for now, support and encourage them to make their decision permanent.

With respect to animal-derived ingredients, I chose to stay with the categories generally used

---

*Mary Kay Cosmetics answered this question with the word "possibly." In a phone conversation to clarify this, they explained their temporary moratorium will possibly change to a permanent policy. They went on to say the company will not return to animal testing.

and accepted. No ingredients derived from animals is designated as category one (1). Honey, beeswax, lanolin, milk, and animal ingredients that do not require the animals to die are designated as category two (2). Animal-derived ingredients that require the death or dismemberment of animals is category three (3). Since the purpose of this book is to help put an end to animal product testing, products that contain animal-derived ingredients were not excluded.

The companies were asked to put in writing whether their products contained animal-derived ingredients. They were also asked to send an ingredient list if they claimed no animal-based ingredients were used. I was going to use these ingredient lists to ensure the accuracy of the ingredient designations. Unfortunately, I found this would not work, as many ingredients that were originally only derived from animals can now be synthesized or made from plants. These synthetic chemicals have the same names as the animal-derived ones. For instance, keratin, used in hair and skin care products, once came only from the hooves of animals, but now it can be derived from human hair. Both forms of keratin have the same name. It was impossible to differentiate animal-derived keratin from human hair keratin by reading the ingredient lists. Luckily, most of the companies marked products that had animal ingredients and told me what ingredient it contained, enabling me to designate each product individually and correctly. I have reported only what they claimed and hope they were honest in their answers.

Some companies did not send ingredient lists and did not specify which products contained animal-derived ingredients, stating only that "some" of their product line did. In an attempt to get correct designations for their products these companies were contacted and given two options. Their first option was to send the correct, individual designations for each product. Unfortunately, this was not practical for some companies with extensive product lines. Some didn't have the time to do this, others didn't have the knowledge. As I had already found out, even if you knew what ingredients to look for, it was impossible to determine the source by reading a label. In some instances, only the chemist who originally formulated the product knew the exact ingredients and their sources.

Therefore, an alternative option was to designate the whole product line as one category, as long as the highest appropriate designation that pertained was used. That is, if they had 20 products with unknown ingredients and one product with a known category three animal by-product ingredient, they had to label *all* their products as containing animal by-products, *even when this may not be true*.

Some companies could not be reached or left it up to me to designate the products. If a company did not designate their products individually but sent ingredient lists for me to make the determination and I found a questionable ingredient on the label, the assumption was made that it was animal-derived and designated as such *even though this might not be correct*. If they did not send ingredient lists and could not be reached in time to clarify the matter, I used their reply to the questionnaire as the designation for the whole line. If their reply was "yes" or "some" concerning the use of animal products, it was assumed they were referring to category three ingredients, and all their products were marked as such. Again, it is conceivable that not all of their products contain these ingredients, but without the ingredient lists, there was no way to be sure. Therefore, I strongly recommend using the ingredient keys as a guide only.

While using this book you may discover categories that are not normally associated with animal testing. When a company that does not test its products on animals is found, all of their products are listed, which results in these odd categories. They were included so that cruelty-free companies can be supported wholeheartedly.

Finally, I would like to say that neither the author nor publisher charged or accepted a fee from any of the companies listed herein and that inclusion in this book in no way constitutes an endorsement of any company or product. To the best of my ability, this book conveys accurate information, although typographical errors, incorrect information supplied by the companies, assumptions in product ingredients (as discussed above), changes in corporate policies, mergers, acquisitions, and other factors may at some point affect the accuracy of this information.

COMPANIES THAT DO NOT TEST ON ANIMALS

* Did not send a product list
** Private labels, no products under this name
*** Bulk supplier to industry only

Abbaco, Inc.
230 Fifth Ave.
New York, NY 10001
212-679-4550

Abkit, Inc.
1160 Park Ave.
New York, NY 10128
212-860-8538

Abracadabra
P.O. Box 1040
Guernerville, CA 95466
800-523-5232; 707-869-0761

Advanced Research Labs
16580 Harbor Blvd., Suite O
Fountain Valley, CA 92708
800-966-6960; 714-839-1940

Adwe Laboratories, Inc.
141 20th St.
Brooklyn, NY 11232
718-788-6838

AFM Enterprises, Inc.
1140 Stacy St.
Riverside, CA 92507
714-781-6860

A.I.I. Clubman
2220 Gaspar Ave.
City of Commerce, CA 90040
800-621-9585; 213-728-2999

A. J. Funk & Co.
147 Timber Dr.
Elgin, IL 60120
708-741-6760

Alba Botanica Cosmetics
P.O. Box 12085
Santa Rosa, CA 95406
707-575-3111

Alexandra Avery
68103 Northrup Creek
Birkenfeld, OR 97016
503-755-2446

Alexandra de Markoff*
625 Madison Ave.
New York, NY 10022
212-572-5000

Allens Naturally
P.O. Box 339, Dept. LC
Farmington, MI 48332-0339
313-453-5410

Alpha 9, JDS Manufacturing
  Co., Inc.
7718 Burnet Ave.
Van Nuys, CA 91405
800-342-5742

American Merfluan
3479 Edison Way
Menlo Park, CA 94025
415-364-6343

Amsource International/
   Royal Laboratories
3780 Selby Ave.
Los Angeles, CA 90034
800-688-4044

Andrea International Ind.
2220 Gaspar Ave.
City of Commerce, CA 90040
800-621-9585; 213-728-2999

Andrew Jergens
P.O. Box 145444
Cincinnati, OH 45250
513-421-1400

Apiana/Bienen Mathys
Forest Road
Marlow, NH 03456
800-327-2324

Aramis, Inc.*
767 Fifth Ave.
New York, NY 10153
212-572-3700

Ardell International, Inc.
2220 Gaspar Ave.
Los Angeles, CA 90040
800-621-9585; 213-728-2999

Arizona Natural Resources,
   Inc.
1331 W. Melinda Lane
Phoenix, AZ 85027
602-869-0801

Aroma Vera
P.O. Box 3609
Culver City, CA 90231
213-675-8219

Aubrey Organics
4419 N. Manhattan Ave.
Tampa, FL 33614
813-877-4186

Aura Cacia
P.O. Box 399
Weaverville, CA 96093
916-623-3301

Auro Trading Company
18A Hangar Way
Watsonville, CA 95076
408-728-4525

Auroma International
Box 2
Wilmot, WI 53192

Auromere/Ayurvedic
1291 Weber St.
Pomona, CA 91768
714-629-8255

Aurora Henna Company
1507 E. Franklin Ave.
Minneapolis, MN 55404
612-870-4456

Autumn Harp
28 Rockydale Road
Bristol, VT 05443
802-453-4807

Avanza Corp./Natures
   Cosmetics/Heavana
881 Alma Real, Suite 118
Pacific Palisades, CA 90272
800-553-9816; 800-433-6290

Aveda Corporation
321 Lincoln St. NE
Minneapolis, MN 55413
400 Central Ave.
Minneapolis, MN 55414
800-328-0849

Avon
9 W. 57th St.
New York, NY 10003
212-546-6015

Baby Touch Ltd.
387 Riviera Cir.
Larkspur, CA 94939
415-924-4060

Barbizon International
950 Third Ave.
New York, NY 10022
212-371-4300

Bare Escentuals/Dolphin
   Acquisition Corp.
809 University Ave.
Los Gatos, CA 95030
408-354-8853

Baudelaire
Forest Road
Marlow, NH 03456
800-327-2324

Beauty Without Cruelty, Ltd.
67A Galli Dr.
Novato, CA 94949
415-382-7784

Beehive Botanicals
Rte. 8, Box 8258
Hayward, WI 54843
800-283-4274

Benetton Cosmetics
540 Madison Ave., 29th Floor
New York, NY 10022
800-722-7070

Best Foods
P.O. Box 1534
Union, NJ 07083-5088
201-688-9000

Betrix/Eurocos*
625 Madison Ave.
New York, NY 10022
212-572-5000

Bio-Botanica, Inc.
75 Commerce Dr.
Hauppage, NY 11788
516-231-5522

Biogime International
1187 Brittore Road
Houston, TX 77043

Biokosma/Dr. Grandel
626 W. Sunset Road
San Antonio, TX 78216
800-543-5230; 512-829-1763

Body Love Natural Cosmetics
P.O. Box 7542
Santa Cruz, CA 95061
408-425-8218

Body Shop (Calif.)
1341 7th St.
Berkeley, CA 94710
415-524-0216

Body Shop International
45 Horsehill Road Tech. Ctr.
Cedar Knolls, NJ 07927-2003
800-541-2535; 201-984-9200

Borlind of Germany
P.O. Box 130
New London, NH 03257
800-447-7024; 603-526-2076

Botanicus Retail
7920 Queenair Dr.
Gaithersburg, MD 20879
800-282-8887

Breezy Balms
Box 168
Davenport, CA 95017
408-423-4299

Bronson Pharmaceuticals
4526 Rinetti Lane
La Canada, CA 91011
800-521-3322

Brookside Soap Co.
P.O. Box 55638
Seattle, WA 98155
206-363-3701

Bug-Off
Rte. 3, Box 27A
Lexington, VA 24450
703-463-1760

C. A. Gregory Aromatics
Forest Road
Marlow, NY 03456
800-327-2324

Carlson Laboratories
15 College Dr.
Arlington Heights, IL 60004
800-323-4141

Carma Laboratories
5801 W. Airways Ave.
Franklin, WI 53132
414-421-7707

Cernitin America, Inc.
100 Corry St.
P.O. Box 839
Yellow Springs, OH 45387
800-831-9505; 513-767-2412

Charles of the Ritz, Ltd.*
625 Madison Ave.
New York, NY 10003
212-572-5000

Chempoint Products Co.
543 Tarrytown Road
White Plains, NY 10607
800-343-6588

Chip Distribution
1139 Dominguez St., Unit E
Carson, CA 90746
213-603-1114

Chuckles Inc.
59 March Ave.
Manchester, NH 03103
603-669-4228

Clear Vue Products, Inc.
417 Canal St.
P.O. Box 567
Lawrence, MA 01842

Clearly Natural /
  Laguna Soap
1306 Dynamic St.
Petaluma, CA 94954
707-762-5815

Clientele
5207 NW. 163rd St.
Miami, FL 33014
800-327-4660

Clinique Laboratories, Inc.*
767 Fifth Ave.
New York, NY 10153
212-572-3800

Colonial Dames Company,
  Ltd.
P.O. Box 22022
Los Angeles, CA 90022
213-773-6441

Color Me Beautiful
14000 E. Thunderbolt Dr.
Chantilly, VA 22021

Columbia Cosmetics**
1661 Timothy Dr.
San Leandro, CA 94577
800-824-3328

Comfort Manufacturing Co.
1056 Van Buren St.
Chicago, IL 60607
312-421-8145

Cosmetic Source, Inc.
i Natural / Natural Solutions
355 Middlesex Ave.
Wilmington, MA 01887
800-9-MAKEUP

Cosmetic Technology
   International
4 Embarcadero Center 5100
San Francisco, CA 94111
408-761-2144

Cosmyl Cosmetics
16115 NW 52nd Ave.
Miami, FL 33014
800-262-4401

Country Comforts
P.O. Box 3
Nuevo, CA 92367
707-584-3057

Crabtree & Evelyn
Peake Brook Road
Box 167
Woodstock, CT 06281
203-928-2761

DeLore Products
2220 Gaspar Ave.
City of Commerce, CA 90040
800-621-9585; 213-728-2999

Dena Corporation / Meta
   International
850 Nicolas Blvd.
Elk Grove Village, IL 60007
708-593-3041

Desert Essence
P.O. Box 588
Topanga, CA 90290
800-727-2714; 213-455-1046

Desert Naturels
1761 Cosmic Way
Glendale, CA 91201
818-247-4970

Dr. Babor Natural Cosmetics
1636 Gervaisa Ave. #9
Maplewood, MN 55109
800-333-4055; 612-770-0611

Dr. E. H. Bronner
P.O. Box 28
Escondido, CA 92025
619-745-7069

Dr. Hauschka Cosmetics
P.O. Box 407
Wyoming, RI 02898
401-539-7037

Duracell USA
Berkshire Industrial Park
Bethel, CT 06801
800-551-2355

Earth Gifts
2424 T St.
Sacramento, CA 95816
916-456-7640

Earth Science
545 Porter Way
Placentia, CA 92670
800-222-6720; 714-524-9277

Ecco Bella
6 Provost Sq. #602
Caldwell, NJ 07006
800-888-5320; 201-226-5799

Ecover Products / Mercantile
   Food Co.
4 Old Mill Road
P.O. Box 1140
Georgetown, CT 06829-1140
203-544-9891

Eden Botanicals
6265 Little Applegate Road
Jacksonville, OR 97530-9303
415-530-3401

Estee Lauder, Inc.*
767 Fifth Ave.
New York, NY 10153
212-572-4200

Eva Jon Cosmetics, Inc.
1016 E. California St.
Gainesville, TX 76240
817-668-7707

Eveready Battery Company
Checkerboard Sq.
St. Louis, MO 63164

The Face Food Shoppe
3704 S. Bethel St.
Columbia, MO 65203
314-443-0895

Faith in Nature
Forest Road
Marlow, NH 03456
800-327-2324

Farmavita
59 March Ave.
Manchester, NHY 03103
800-221-3496

Fashion and Designer
  Fragrances, Inc.*
625 Madison Ave.
New York, NY 10022
212-572-5000

Finelle Cosmetics
137 Marston St.
P.O. Box 5200
Lawrence, MA 01842
617-682-6112

Focus 21 International, Inc.
2755 Dos Aarons Way
Vista, CA 92083
619-727-6626

Forever New
Sioux Falls, SD 57101

G. T. International
1800 S. Robertson Blvd.,
  #182
Los Angeles, CA 90035
800-359-2940; 213-551-0484

Germaine Monteil*
625 Madison Ave.
New York, NY 10022
212-572-5000

Golden Lotus/Mountain
  Fresh
P.O. Box 40516
Grand Junction, CO 81504
303-434-8434

Granny's Old Fashioned
  Products
P.O. Box 256
Arcadia, CA 91006
818-577-1825

Green Ban
Box 146
Norway, IA 52318

Gruene Kosmetik
256 S. Robertson Blvd.
Box 4122
Beverly Hills, CA 90211
213-392-2449

H. A. Cole Products Company
P.O. Box 9937
Jackson, MS 39286-0937
601-366-9325

Hasbro Toys
1027 Newport Ave.
P.O. Box 1059
Pawtuket, RI 02861-1059
800-242-7276

Heavenly Soap
5948 E. 30th St.
Tucson, AZ 85711
602-790-9938

Heritage Store
P.O. Box 444
Virginia Beach, VA 23458
804-428-0100

Hobe Laboratories, Inc.
201 S. McKemy Ave.
Chandler, AZ 85226
602-257-1950

Hofels
Forest Road
Marlow, NH 03456
800-327-2324

Holloway House, Inc.
8328 Masters Road
P.O. Box 50126
Indianapolis, IN 46250
800-255-1891

Home Health Products, Inc.
P.O. Box 3130
Virginia Beach, VA 23454
800-468-7313; 804-491-2200

Home Service Products
P.O. Box 269
Bound Brook, NJ 08805
201-356-8175; 215-348-2393

House of Lowell, Inc.
1172 W. Galbraith Road
Cincinnati, OH 45231
513-521-2212

Hubner Bath Oils
Forest Road
Marlow, NH 03456
800-327-2324

Iced Creme Facial Masque
1001 Bridgeway, Suite 534
Sausalito, CA 94965
800-548-6999

Il-Makiage
107 E. 60th St.
New York, NY 10022
800-722-1011; 718-361-3123

Indian Creek
P.O. Box 63
Selma, OR 97538
503-592-2616

Institute of Trichology
1619 Reed St.
Lakewood, CO 80215
303-232-6149

Integrated Health
12832 Chadron Ave.
Hawthorne, CA 90250
800-367-7744; 213-675-1080

International Rotex
P.O. Box 20697
Reno, NV 89515
702-356-8356

Irma Shorrell, Inc.
P.O. Box 617
Madison Heights, VA 24572
800-523-0907

Jacki's Magic Lotion
258 A St. #7A
Ashland, OR 97520
503-488-1388

Jaclyn Cares
P.O. Box 339, Dept. LC
Farmington, MI 48332-0339
313-453-5410

Jacob Hooy & Co.
Forest Road
Marlow, NY 03456
800-327-2324

J & L Products
P.O. Box 63241
Phoenix, AZ 85082
800-666-0373

Jason Natural Products
8468 Warner Dr.
Culver City, CA 90232
213-838-7543

Jean Pax, Inc.
P.O. Box 560475
Miami, Fl 33156
800-327-1968; 305-593-0982

Jeanne Gatineau*
625 Madison Ave.
New York, NY 10022
212-572-5000

Jeanne Rose Herbal Body
  Works
219-A Carl St.
San Francisco, CA 94117
415-564-6785

Jelene International, Inc.
1409 Allen Dr., Suite G
Troy, MI 48083
313-588-5522

John F. Amico & Co.*
7327 W. 90th St.
Bridgeview IL 60455
708-430-2552

John Paul Mitchell Systems
P.O. Box 10597
Beverly Hills, CA 90213-3597
800-321-5767

JOICO Laboratories
345 Baldwin Park Blvd.
City of Industry, CA 91746

Jolen, Inc.
25 Walls Dr.
Fairfield, CT 06430
203-259-8779

J. R. Liggett, Ltd.
RR 2, Box 911, Rte. 12-A
Cornish, NH 03745
603-675-2055

Jurlique Cosmetics/
  D'Namis, Ltd.
16 Starlit Dr.
Northport, NY 11768
800-642-0666; 516-754-3535

Kallima International Inc.
10500-122 Metric Dr.
Dallas, TX 75243
214-553-5824

Kenra Laboratories
6501 Julian Ave.
Indianapolis, IN 46219
800-428-8073; 317-356-6491

Key West Fragrance &
  Cosmetic Factory
P.O. Box 1079
524 Front St.
Key West, FL 33041
800-445-2563; 800-433-2563

Kids William & Clarissa
23041 Avenida Carlota, Suite
  325
Laguna Hills, CA 92653
800 722 5437; 714 770 4001

Kimberly Sayer, Inc.
61 W. 82nd St.
New York, NY 10024

Kiss My Face
P.O. Box 224
Gardiner, NY 12525
914-255-0884

Kiwi Brands
Rt. 662
Douglasville, PA 19518
215-385 3041

KMS Research, Inc.
6807 Hwy. 299 E.
P.O. Box 520
Bella Vista, CA 96008
916-549-4472

KSA Jojoba
19025 Parthenia St. #200
Northridge, CA 91324
818-701-1534

La Costa Products
2251 Las Palmas Dr.
Carlsbad, CA 92009-4878
619-438-2181

La Crista, Inc.
P.O. Box 240
Davidsonville, MD 20135
301-956-4447

Lady of the Lake Company
6351 Wilshire Blvd. #212
Los Angeles, CA 90048
213-837-2954

Lakon Herbals
4710 Templeton Road
Montpelier, VT 05602
802-223-5563

L'anza Research Laboratories
5523 Avon Ave.
Irwindale, CA 91706
800-423-0307; 818-334-9333

L'Arome (USA), Inc.
456 Lakeshore Pkwy.
Rock Hill, SC 29730
800-798-3129; 803-329-3756

Levlad/Nature's Gate/
   Alogen
9183-5 Kelvin Ave.
Chatsworth, CA 91311
800-327-2012; 818-882-2951

Liberty Natural Products
P.O. Box 66068
Portland, OR 97266
800-289-8427

Lily of Colorado
1286 S. Valentia St.
Denver, CO 80231
303-455-4194

Livos Plant Chemistry
1365 Rufina Cir.
Santa Fe, NM 87501
800-621-2591; 505-988-9111

Lotus Light/Light Touch
P.O. Box 2
Wilmot, WI 53192
414-862-2395

Luzier Personalized
   Cosmetics*
3216 Gilham Plaza
Kansas City, MO 64109
816-531-8338

Magic American Corp.
23700 Mercantile Road
Cleveland, OH 44122
800-321-6330; 216-464-2353

Magic of Aloe
7300 N. Crescent Blvd.
Pennsauken, NJ 08110
800-257-7770; 609-662-3334

Marchemco Cosmetics Co.
Box 1010
Bronxville, NY 10708
203-838-5515; 914-793-2093

Mary Kay Cosmetics
8787 Stemmons Pkwy.
Dallas TX 75247
214-630-8787

Mary Quant*
625 Madison Ave.
New York, NY 10022
212-572-5000

Mastey de Paris
25413 Rye Canyon Road
Valencia, CA 91355
805-257-4814

Max Factor*
625 Madison Ave.
New York, NY 10022
212-572-5000

Mellita USA, Inc.
1401 Berlin Road
Cherry Hill, NJ 08003
609-427-2803

Mera Corp.
401 N. 3d St. #300, Designers
  Guild Bldg.
Minneapolis, MN 55401
800-752-7261; 612-332-2136

Mia Rose Products, Inc.
1374 Logan, Unit C
Costa Mesa, CA 92626
714-662-5891

Michael's Health Products
5085 List Dr. #100
Colorado Springs, CO 80919
512-647-4700

Michel Constantini Natural
  Cosmetics
124 W. 72nd St.
New York, NY 10023
800-321-5012; 212-799-3442

Micro Balanced Products
  Corp.
25 Aladdin Ave.
Dumont NJ 07628
201-387-0200

Milton Bradley Toys
1027 Newport Ave.
Pawtucket, RI 02862-1059
401-726-4100

Miner's*
625 Madison Ave.
New York, NY 10022
212-572-5000

Mountain Ocean
P.O. Box 951
Boulder, CO 80306
303-444-2781

Murphy-Phoenix Company
P.O. Box 22930
Beachwood, OH 44122
800-486-7627; 216-831-0404

Natural Touch
12527 130th Lane NE
Kirkland, WA 98034
206-820-2788

Naturally Yours Alex
P.O. Box 2365
Pleasant Hill, CA 94523
415-798-2520

Nature De France
444 Park Ave. S
New York, NY 10016
800-237-9418; 212-213-4343

Nature's Colors/Ida Grae
  Cosmetics
424 Laverne Ave.
Mill Valley, CA 94941
415-388-6101

NatureWorks, Inc.
5341 Derry Ave. #F
Agoura Hills, CA 91301
818-889-1602

Nectarine**
1200 5th St.
Berkeley, CA 94710
415-528-0162

New Beginnings Natural
    Bodycare
511 Cale
San Pablo, Camarillo, CA
    93010
805-482-7791

New Moon Extracts/
    Geremy Rose Cosmetics
105 Main St.
P.O. Box 1947
Brattleboro, VT 05301
800-543-7279; 802-257-0018

Neway
Little Harbor
Marblehead, MA 01945
617-227-5117

Nirvana
P.O. Box 18413
Minneapolis, MN 55418
612-932-2919

No Comment Scents
Kings Yard
Yellow Springs, OH 45387
513-767-4260

North Country Naturals, Inc.
1301 39th St. N
Fargo, ND 58102
701-282-2300

North Country Soap
7888 County Road 6
Maple Plain, MN 55359
800-328-4827 ext 2153;
612-479-3381

NuSun/Marchemco
Box 1010
Bronxville, NY 10708
203-838-5515; 914-793-2093

Nutri-Metrics International,
    Inc.
19501 E. Walnut Dr.
P.O. Box 128
City of Industry, CA 91749
714-598-1831

O'Naturel
P.O. Box 70504
Oakland, CA 94607
415-268-9121

Oriflame International
76 Trebel Cove Road
N. Billerica, Ma 01862
508-663-2700

Origins Natural Resources,
    Inc.*
767 Fifth Ave.
New York, NY 10153
212-572-4100

Orjene Natural Cosmetics
5-43 48th Ave.
Long Island City, NY 11101
718-937-2666

Oxyfresh U.S.A.
10906 Marietta
Spokane, WA 99206
509-924-4999

Patricia Allison
4470 Monahan Road
Le Mesa, CA 92041
619-444-4879

Paul Mazzotta
P.O. Box 96
Reading, PA 19607
215-376-2250

Paul Penders USA
1340 Commerce St.
Petaluma, CA 94952
707-763-5828

Peelu Products, Ltd.
6224 Madison Ct.
Morton Grove, IL 60053
708-967-9400

Pet Connection
P.O. Box 391806-BB
Mountain View, CA 94039
415-949-1190

PetGuard
165 Industrial Loop S, Unit 5
Orange Park, FL 32073
800-874-3221; 800-331-7527

Pets 'N People, Inc.
5312 Ironwood St.
Rancho Palos Verdes, CA
   90274
213-373-1559

Phebo
Forest Road
Marlow, NH 03456
800-327-2324

Phillip Rockley, Ltd.***
20505 Dag Hammarskjöld
   Conv. Center
New York, NY 10017
212-355-5770

Playskool Toys
1027 Newport Ave.
Pawtucket, RI 02862-1059
401-726-4100

Potions & Lotions
10201 N. 21 Ave. #8
Phoenix, AZ 85021
602-944-6642; 602-272-3042

Prescriptives, Inc.*
767 Fifth Ave.
New York, NY 10153
212-572-4400

Prestige Fragrances, Ltd.*
625 Madison Ave.
New York, NY 10022
212-572-5000

PRH & Associates, Inc.
P.O. Box 2741
Peachtree City, GA 30269
800-451-7096

Princess Marcella Borghese*
625 Madison Ave.
New York, NY 10022
212-572-5000

Pro-File
2220 Gaspar Ave.
City of Commerce, CA 90040
800-621-9585; 213-728-2999

Pro-Ma Systems
976 Florida Central Pkwy.
   #136
Longwood, FL 32750
407-331-1133

Professional & Technical
   Services, Inc.
3331 NE. Sandy Blvd.
Portland, OR 97232-1926
800-648-8211; 503-231-7244

Puritans Pride
105 Orville Dr.
Bohemia, NY 11716
516-567-9500

Rachel Perry, Inc.
9111 Mason Ave.
Chatsworth, CA 91311
800-624-7001; 818-888-5881

Rainbow Research
170 Wilbur Pl.
Bohemia, NY 11716
800-722-9595; 516-589-5563

Rathdowny
3 River St.
P.O. Box 357
Bethel, VT 05032
800-543-8885; 802-234-9928

Reviva Labs
705 Hopkins Road
Haddonsfield, NJ 08033
800-257-7774

Revlon*
625 Madison Ave.
New York, NY 10022
212-572-5000

SafeBrands, Inc.
55 W. Sierra Madre Blvd.
Sierra Madre, CA 91024
818-355-1050

Sappo Hill Soap Works
654 Tolman Creek Road
Ashland, OR 97520
503-482-4485

Sarakan
Forest Road
Marlow, NH 03456
800-327-2324

Scarborough and Company
P.O. Box 637
Wilton, NH 03086
603-673-3800

Sea Minerals Co.
535 Fifth Ave., Suit 809
New York, NY 10017
800-538-8383; 212-286-9290

Sebastian International
6109 Desoto Ave.
Woodland Hills, CA 91367
818-712-7700

SerVaas Laboratories, Inc.
1200 Waterway Blvd.
P.O. Box 7
Indianapolis, Indiana 46207
317-636-7760

Seventh Generation
49 Hercules Dr.
Colchester, VT 05446-1672
800-441-2538; 800-456-1177

Shahin Soap Co.
P.O. Box 8117
Haledon, NJ 07508
201-790-4296

Shaklee U.S., Inc.
444 Market St.
San Francisco, CA 94111
800-Shaklee (742-5533)

ShiKai Products/Trans India
P.O. Box 2866
Santa Rosa, CA 95404
800-448-0298; 707-584-0298

Shirley Price Aromatherapy
462 62nd St.
Brooklyn, NY 11220
718-492-3887

Sierra Dawn
P.O. Box 513
Graton, CA 95444

Similasan Corp.
23224 94th Ave. S
Kent, WA 98031

Simplers Botanical Co.
Box 39
Forestville, CA 95436
707-887-2012

Sirena Soap/Tropical Soap
   Co.
P.O. Box 831080
Richardson, TX 75083-1080
800-527-2368; 214-243-1991

Smith & Vandiver, Inc.
480 Airport Blvd.
Watsonville, CA 95076
408-722-9526

Soap Factory
141 Cushman Road
St. Catharines, ONT L2M
   6T2, Canada
416-682-1808

SoapBerry Shop*
50 Galaxy Blvd., Unit 12
Rexdale, Ontario M9W 4Y5,
   Canada
800-387-4818; 416-674-0248

Solgar Vitamin Co.
410 Ocean Ave.
Lynbrook, NY 11563
718-895-2612; 516-599-2442

Solid Gold Holistic Animal
   Equine Nutrition*
1483 N. Cuyumaca
El Cajon, CA 92020
619-465-9507

Sombra/C & S Laboratories
5600G McLeod Ave.
Albuquerque, NM 87109
800-225-3963; 505-888-0288

Sterns Miracle Grow
800 Port Washington Blvd.
Port Washington, NY 11050
800-645-8166

Strong Skin Savvy, Inc.
4 Lakeside Dr.
New Providence, PA 17560
717-786-8947

Studio Magic
1417-2 Del Prado Blvd. #480
Cape Coral, FL 33990
813-283-5000

Sukesha
59 March Ave.
Manchester, NH 03103
800-221-3496

Sunshine Products Group
1919 Burnside Ave.
Los Angeles, CA 90016
213-929-6400

Super Nail
2220 Gaspar Ave.
City of Commerce CA 90040
800-621-9585; 213-728-2999

Surrey, Inc.
13110 Trails End Road
Leander, TX 78641
512-267-7172

TerraNova
1200 5th St.
Berkeley, CA 94710
415-528-0666

TLC Pet Products, Inc.
33317 W. Washington Blvd.
Los Angeles, CA 90018
213-399-7146; 213-730-1690

Tom's of Maine
P.O. Box 710, Railroad Ave.
Kennebunk, ME 04043
207-985-2944

Touchstone Company
P.O. Box 686
Wrightsville, NC 28480

Tyra Skin Care
9010 Oso Ave., Unit A
Chatsworth, CA 91311

Ultima II*
625 Madison Ave.
New York, NY 10022
212-572-5000

Venus Laboratories, Inc.
855 Lively Blvd.
Wood Dale, IL 60191
800-624-1900; 800-592-1900

Visage Beaute*
625 Madison Ave.
New York, NY 10022
212-572-5000

Wachters Organic Sea
   Products
360 Shaw Road S
San Francisco, CA 94080

Warm Earth Cosmetics
2230 Normal Ave.
Chico, CA 95928
916-895-0455

Weleda, Inc.
P.O. Box 769, Dept. LC
Spring Valley, NY 10977
914-356-4134

Wilkinson Sword
7012 Best Friend Dr.
Atlanta, GA 30340
800-241-8918; 800-441-3034

WiseWays Herbals
Box 50
Montague, MA 01351
413-625-9509

Wite-Out Products
10114 Bacon Dr.
Beltsville, MD 20705-2197
800-638-0994; 301-937-5353

Withers Mill Co.
P.O. Box 347
Hannibal, MO 63401
800-223-0858; 314-221-4747

World of Aloe
P.O. Box 821635
Ft. Worth, TX 76182
800-522-2563 (ALOE)

W. T. Rawleigh Company
3730 W. Windover
Greensboro, NC 27407
800-525-6050

Wysong Corp.
1880 N. Eastman Road
Midland, MI 48640
517-631-0009

Xavier Distributors
10 Union Hall Ct.
Baltimore, MD 21228
800-543-4379

Zia Cosmetics
300 Brannan St., Suite 601
San Francisco, CA 94107
800-334-7546; 415-543-SKIN

Zinzare International
11308 Hartland St.
N. Hollywood, CA 91605
800-544-6880; 818-760-3702

CRUELTY-FREE CATALOGS

ABEnterprises
P.O. Box 120220
Staten Island, NY 10312-0006
718-948-2622

Amberwood
Rt. 1, Box 206
Milneer, CA 30257
404-358-2991

Atlantic Recycled Paper
   Company
P.O. Box 11021
Baltimore, MD 21212
301-323-2676

Basically Natural
109 E. G St.
Brunswick, MD 21716
301-834-7923

Baubiologie Hardware
207-B 16th St.
Pacific Grove, CA 93950
408-372-8626

Baudelaire
Forest Road
Marlow, NH 03456
800-327-2324; 603-352-9234

Beauty Naturally
57 Bosque Road
P.O. Box 429
Fairfax, CA 94930
415-459-2826

Blue Ribbons Pet Care
2475 Bellmore Ave.
Bellmore, NY 11710
516-785-0604

Carole's Cosmetics
7678 Sagewood Dr.
Huntington Beach, CA 92648
714-842-0454

Compassionate Consumer
Box 27
Jericho, NY 11753
718-445-4134

Earth Care Paper, Inc.
P.O. Box 14140
Madison, WI 53714-0140
608-277-2900

Earthsafe Products
P.O. Box 81061
Cleveland, OH 44181-9933
216-734-1230

Earthwise Products
P.O. Box 683
Roseville, MI 48066-0683
313-772-6999

Eco-Choice
P.O. Box 281, Dept. 5060
Montvale, NJ 07645-0281
800-535-6304; 201-930-9046

Eco-Design/Livos
1365 Rufina Cir.
Santa Fe, NM 87501
505-438-3448

Everybody, Ltd.
1738 Pearl St.
Boulder, CO 80302
303-440-0188

Feather River Company
133 Copeland
Petaluma, CA 94952
800-762-8873

Heart's Desire
1307 Dwight Way
Berkeley, CA 94702

Home Health Products
P.O. Box 3130
Virginia Beach, VA 23454
800-284-9123

Humane Alternative Products
8 Hutchins St.
Concord, NH 03301
603-224-1361

If You Love Animals, Inc.
P.O. Box 409
Linden, MI 48451
313-750-1776

InterNatural
Shaker St.
P.O. Box 680
S. Sutton, NH 03273
800-446-4903

Kindness Products
7150 Morse Road
New Albany, OH 43054
614-855-1895

Lion & Lamb
29–28 41st Ave.
Lond Island City, NY 11101
800-252-6288; 718-361-5757

Naturally Ewe
14622 SR 574 W
Dover, FL 33527
813-681-6787

Nature's Choice
P.O. Box 18381
Pittsburgh, PA 15236
412-391-0988

Painlessly Beautiful
1260 Lumber St.
Middletown, PA 17057
717-939-5376

Peaceable Kingdom
1902 W. 6th St.
Wilmington, DE 19805

Rainbow Concepts
Rte. 5, Box 569-H,
Pleasant Mtn.
Toccoa, GA 30577
404-886-6320

Rare Earth Enterprises
P.O. Box 172
Wilmot, WI 53192
414-862-6347

Seventh Generation
49 Hercules Dr.
Colchester, VT 05446-1672
800-441-2538

Spare the Animals
P.O. Box 233-LC
Tiverton, RI 02878
401-625-5963

Sunrise Lane Products
780 Greenwich St.
New York, NY 10014
212-242-7014

Super Nature
4 Civic Pl.
Bloomfield, NJ 07003
201-338-6026

Withers Mill Company
P.O. Box 347

Hannibal, MO 63401
800-223-0858

Without Harm
4605 Pauli Dr.
Manlius, NY 13104
315-682-8346

Yvonne Richards, Ltd.
P.O. Box 476
Whitehouse, NJ 08888

The combined force of a few thousand sparks makes a powerful bolt of lightning.

Arlo Guthrie

No army can withstand the strength of an idea whose time has come.

Victor Hugo

A SHOPPER'S GUIDE TO
**ALPHABETICAL
LISTING**
CRUELTY-FREE PRODUCTS

KEY TO THE LISTS

The products on the following pages are grouped according to where they are primarily sold. Many companies sell their products in more than one location. This is to be used as a guide only. If you are having problems locating a product, please contact the manufacturer. When contacting a firm in writing, it is best to include a self-addressed, stamped envelope.

General Retail Stores—Products with this heading can be found in mass market retail stores, such as supermarkets, drug, department, or gift stores.

Hair Salons—This category designates products that can be found in hair and skin salons or beauty supply stores.

Health Food Stores—Companies whose products are sold in health and natural food stores will be under this listing.

Home Shopping/Mail Order—This category designates companies whose products can be bought without leaving your home. They may be found in catalogs, purchased by mail or phone, or ordered through company representatives who come into your home.

Specialty Stores—This designates products found in company owned and operated stores, chains, franchises.

The number following each product and company name designates whether or not the product contains animal-derived ingredients. Many companies did not designate each product individually but rather assigned the highest number designation that pertained to the line as a whole. This is explained further in "Guidelines for This Book." Therefore, use these numbers as a guide only.

1—This number designates that no animal-derived ingredients are used in the product.

2—The products marked with this number contain animal-derived ingredients that do not require the death of the animal, such as honey, beeswax, milk, eggs, or lanolin.

3—This designates products that contain animal ingredients that required the death or dismemberment of the animal. Many ingredients that once came only from this category can now be found in plants, or synthesized chemically. Do not rely on the ingredient name to determine their origins.

A

ACNE CARE PRODUCTS

General Retail Stores

Blemish Zapper—Color Me Beautiful, 1
Ice-O-Derm Deep Cleansing Bar—A.I.I. Clubman, 1
Ice-O-Derm Medicated Astringent Gel—A.I.I.
  Clubman, 1
Ice-O-Derm Medicated Astringent Gel—A.I.I.
  Clubman, 1

Health Food Stores

Aminoderm Gel—Aubrey Organics, 1
Blackmores Thyme Lotion—Solgar Vitamin Co., 1
Dermaglow—Heritage Store, 1
Homeopathic Acne Cream—Bio-Botanica, 1
Oil Control Extract—Zia Cosmetics, 1

Home Shopping/Mail Order

Acne Treatment Gel—Mary Kay Cosmetics, Inc., 3
Acne Wash—The Face Food Shoppe, 1
Aloe Blemish Creme—Magic of Aloe, 3
Aloe Drawing Lotion—Magic of Aloe, 3
Blemish Control Toner—Mary Kay Cosmetics Inc., 3
Blemish Solver—Avon Products, Inc., 3
BPO Cleanser—Michel Constantini Natural
  Cosmetics, 2
BPO Gel—Michel Constantini Natural Cosmetics, 2
BPO Scrub—Michel Constantini Natural
  Cosmetics, 2

Jojoba Acne Medication—KSA Jojoba, 1
Men's Blemish Control Formula—Mary Kay
    Cosmetics, Inc., 3
Multi-Active Clearing Creme—Kallima
    International, Inc., 1
Teen Skin Pads—Hobe Laboratories, Inc., 1
Zit-Zapper—The Face Food Shoppe, 1

Specialty Stores _____

Camphor Lotion—Potions & Lotions, 3

ADHESIVE PRODUCTS

Home Shopping/Mail Order _____

3 in 1 Adhesive—AFM Enterprises, Inc., 1
Carpet Adhesive—AFM Enterprises, Inc., 1
Lavo-Wallpaper Paste—Livos Plant Chemistry, 1
Linami-Cork Adhesive—Livos Plant Chemistry, 1
Wallpaper Adhesive—AFM Enterprises, Inc.,1

ADHESIVE TAPES

General Retail Stores _____

Book and Reinforcement Tapes—International
    Rotex, 1
Colored Cloth Tape—International Rotex, 1
Colored Plastic Tape—International Rotex, 1
Duct Tape—International Rotex, 1
Masking Tape—International Rotex, 1
Package Sealing Tape—International Rotex, 1
Strapping Tape—International Rotex, 1

AFTER BATH SPLASHES AND OILS

General Retail Stores

China Rose After Bath Oil—TerraNova, 1
De Maris After Bath Splash—Cosmyl Cosmetics, 3
Pikaki After Bath Oil—TerraNova, 1
Tropical Gardenia After Bath Oil—TerraNova, 1

Health Food Stores

After Bath Body Splash—Nature's Gate, 1
Keoki Papaya Bath Oil Spray—Natural Solutions, 1
Lavender Water—Heritage Store, 1
Moisture Plus Protective After Bath Mist—Earth Science, 1

Home Shopping/Mail Order

After Bath Lotion—Shirley Price Aromatherapy, 1
Jojoba Spray Mist—KSA Jojoba, 1

AFTERSHAVES AND COLOGNES

General Retail Stores

Aloe Vera Aftershave—TerraNova, 3
Citrus Musk Colognes—A.I.I. Clubman, 2
Clubman After Shave Lotions—A.I.I. Clubman, 2
Desert Rain Cologne—TerraNova, 3
Eau de Portugal—A.I.I. Clubman, 2
Eau de Quinine—A.I.I. Clubman, 2
Fleur de France Cologne—A.I.I. Clubman, 2
Lilac Vegetal—A.I.I. Clubman, 2
Lime Sec Cologne—A.I.I. Clubman, 2
Naturelle Cologne—A.I.I. Clubman, 2
Vitamin Fortified Soothing Aftershave—TerraNova, 3

Health Food Stores _____

After Shave Formula for Problem Skin—Earth
   Science, 1
After Shave Splash—Alba Botanica, 1
After Shave Tonic—Jason Natural Products, 1
Ginseng Mint Aftershave—Aubrey Organics, 1

Home Shopping/Mail Order _____

Aftershave—Body Shop, 2
Aftershave Gel—Key West Fragrance & Cosmetic
   Factory, Inc., 1
Aloe Aftershave—Gruene, 1
Bamboo—Key West Fragrance & Cosmetic
   Factory, Inc., 1
Cayo Hueso—Key West Fragrance & Cosmetic
   Factory, Inc., 1
Espionage Cologne—Key West Fragrance &
   Cosmetic Factory, Inc., 1
Key West Gold—Key West Fragrance & Cosmetic
   Factory, Inc., 1
L'Arome Echoes Aftershave/Cologne—L'Arome
   USA, Inc., 1
Matchmates—Key West Fragrance & Cosmetic
   Factory, Inc., 1
Menstuff—Key West Fragrance & Cosmetic
   Factory, Inc., 1
Metro Men's Conditiong Aftershave—Shaklee U.S.,
   Inc., 1
Mr. Aloe After Shave—Magic of Aloe, 3
1001 Portholes—Key West Fragrance & Cosmetic
   Factory, Inc., 1
Protective Splash—Biogime, 1
Romano Aftershave Splash—Key West Fragrance
   & Cosmetic Factory, Inc., 1
Sexy Afternoon—Key West Fragrance & Cosmetic
   Factory, Inc., 1
Tioga Men's Cologne—Shaklee U.S., Inc., 1

Zarino Aftershave and Cologne—Key West
    Fragrance & Cosmetic Factory, Inc., 1

Specialty Stores ───────────────────

After Shave Lotion—Potions & Lotions, 3
Aftershave Lotion—Body Shop (Calif.), 3
Designer Reproductions—Potions & Lotions, 3
Men's Range Aftershave Sprinkle—Crabtree &
    Evelyn, Ltd., 1
Men's Range Cologne Spray—Crabtree & Evelyn,
    Ltd., 1

AIR FRESHENERS AND ODOR ELIMINATORS

General Retail Stores ───────────────────

Citressence Air Refresher—Wysong Corp., 1
Citressence Concentrate—Wysong Corp., 1
Citressence Drops—Wysong Corp., 1
Citressence Industrial Wick—Wysong Corp., 1
Mother's Little Miracle—Pets 'N People, 1
Nature's Miracle Air Freshener Deodorant—Pets 'N
    People, 1
Nature's Miracle Stain & Odor Remover—Pets 'N
    People, 1
Smith & Valentine Spray Sachet—Smith &
    Vandiver, Inc., 1
Unifresh-Liquid Odor Eliminating Concentrate—
    Venus Laboratories, Inc., 1

Health Food Stores ───────────────────

Air Fresheners Spray Mist—Liberty Natural
    Products, 1
Air Therapy—Mia Rose Products, 1
Air-O-Ma Air Freshners—Auroma International, 1
All Purpose Deodorizer—Oxyfresh, 1
Citrus Magic—PRH & Associates, Inc., 1
Citrus II—Xavier, 1

Home Shopping/Mail Order

Ecco Bella Citrus Purifying Mist—Ecco Bella, 1
Purifying Mist—Ecco Bella, 1

Specialty Stores

Room Sprays—Crabtree & Evelyn, Ltd., 1
Room Sprays—Scarborough and Company, 1

ALL-PURPOSE CLEANERS

General Retail Stores

All Purpose Spray Cleaner—Soap Factory, 1
Citra-Solve Citrus Solvent—Chempoint Products, 1
*Earth Friendly Natural Orange Concentrated All
    Purpose Cleaner*—Venus Labs, Inc., 1
Earth Friendly Neutral Cleaner— Venus Labs,
    Inc., 1
Earth Rite All-Purpose Cleaner—Magic American
    Corp., 1
Fyne-Pyne—H. A. Cole Products, 1
Household Cleaner—Soap Factory, 1
Kleen All-Purpose Cleaner—Mountain Fresh
    Products, 1
Murphy's Oil Soap—Murphy-Phoenix Company, 1
Pine Plus—H.A. Cole Products, 1

Hair Salons

Blue Gem All-Purpose Cleaner—House of Lowell,
    Inc., 3
Jurlique's Sparkle—Jurlique D'Namis, Ltd., 1

Health Food Stores

Citri-Shine—Mia Rose Products, 1

Citrus Organic Cleaner—Natural Bodycare, 1
Cream Cleanser—Ecover, 1
Golden Lotus All Purpose Cleaner—Lotus Light, 1
Home-Soap Household Cleaner—Sierra Dawn, 1
New America All-Purpose Cleaner—Abkit, Inc., 1
Orange Magic—Xavier, 1
Sal Suds Organic Cleanser—Dr. Bronner's, 1

Home Shopping/Mail Order

Allens All-Purpose Cleaner—Allens Naturally, 1
Allens Multipurpose Spray Cleaner—Allens
   Naturally, 1
AVI Soap Concentrate—Livos Plant Chemestry, 2
Avi-Soap Concentrate—Livos Plant Chemistry, 1
Basic-H—Shaklee U.S., Inc., 1
Con-Lei—CHIP Distribution, 1
Cream Cleaner—Seventh Generation, 1
Latis-Natural Soap—Livos Plant Chemistry, 1
Liquid At-Ease—Shaklee U.S., Inc., 1
Professional All-Purpose Spray Cleaner—Home
   Service Products, 1
Safety Clean—AFM Enterprises, Inc., 3
Spra-Itt—Wachter's, 1
Super Clean—AFM Enterprises, Inc., 3
Uni-Kleen—Wachter's, 1

Specialty Stores

Bio-Clean—Body Shop (Calif.), 1

ALOE VERA

General Retail Stores

Aloe Vera Gel—Colonial Dames Co., Ltd., 1

Pure Gold 90% Concentrated Aloe Vera Liniment—
Mountain Fresh Products, 1
Pure Gold 100% Natural Aloe Vera Gel—Mountain
Fresh Products, 1

Hair Salons

Aloe Vera Gel—Jelene International, 1
Super Naturals Aloe Vera Skin Gel—North Country
Naturals, 2

Health Food Stores

Aloe Concentrate with Herbs—Eva Jon Cosmetics, 3
Aloe 84% Creme w/Vitamin E—Jason Natural
Products, 1
Aloe Vera Gel—Orjene Natural Cosmetics, 3
Aloe Vera Lotion—Amsource International/
Royal Labs, 1
Aloe Vera 98% Gel Moisturizer w/Spiralina—Jason
Natural Products, 1
Aloe Vera 70% Creme—Jason Natural Products, 1
Arizona Naturals 100% Aloe Vera Soothing Gel—
Arizona Natural Resources, Inc., 1
Eva Jon Jelly—Eva Jon Cosmetics, 3
100% Pure Aloe Vera Gel—Aubrey Organics, 1
Out of the Leaf Aloe Vera Fillet—Aubrey Organics, 1

Home Shopping/Mail Order

Active-Aloe Soothing Gel—Kallima International,
Inc., 1
Aloe Aid—Magic of Aloe, 3
Aloe Crystal Gel—Magic of Aloe, 3
Aloe Gel—Body Shop, 2
Aloe Gel—World of Aloe, 1
Aloe Magic Spray—Magic of Aloe, 3
Aloe Vera Gel Plus Herbs—Wachter's, 1
Aloemax 100—Key West Fragrance & Cosmetic
Factory, Inc., 1

Aloe-Thera with Vitamin E—Key West Fragrance
    & Cosmetic Factory, Inc., 2
Mrs. Ewald's Original Gel—Magic of Aloe, 3

Specialty Stores _____

Aloe Vera Gel—Body Shop (Calif.), 1

ANTIPERSPIRANTS AND DEODORANTS

General Retail Stores _____

Le Stick Green Clay Deodorant—Nature de
    France, 3
Le Stick Rose Clay Deodorant—Nature de France, 3
Le Stick White Clay Deodorant—Nature de France, 3
Natural Anti-Perspirant Deodorant—Tom's of
    Maine, 1
Natural Deodorant Roll-on—Tom's of Maine, 1
Natural Deodorant Stick—Tom's of Maine, 1
Wysong Deodorant—Wysong Corp., 1

Hair Salons _____

Super Naturals Aloe Vera Roll-On Deodorant—
    North Country Naturals, 2
Thai Crystal Deodorant Stone—Studio Magic, 2

Health Food Stores _____

Aloe Vera Roll-On Deodorant—Jason Natural
    Products, 1
Aloe Vera Stick Deodorant—Jason Natural
    Products, 1
Aurelius Deodorant for Men—Beauty Without
    Cruelty, Ltd., 1
Calendula Blossom Anti-Perspirant—Aubrey
    Organics, 1

Citrus Deodorant—Weleda, Inc., 1
Deodorant Crystal—Heritage Store, 1
E Plus High C Roll On Deodorant—Aubrey
Organics, 1
Lavilin Deodorant—Micro-Balanced Products
Corp., 1
Madonna Lily Deodorant—Beauty Without Cruelty,
Ltd., 1
Mineral Salts—J & L Products, 1
Natural Dry Pine Deodorant—Aubrey Organics, 1
Natural Sage Deodorant—Weleda, Inc., 1
Natural Scent True Stick Deodorants—Orjene
Natural Cosmetics, 1
*Nature's Gate Herbal Fresh Natural Stick
Deodorant*—Levlad, 1
Nature's Gate Natural Roll-On Deodorant—
Levlad, 1
Oxyfresh Natural Deodorant—Oxyfresh, 1
Tea Tree Deodorant—Desert Essence, 1
Yolanda Deodorant—Beauty Without Cruelty,
Ltd., 1

Home Shopping/Mail Order

Aloe Roll-On Deodorant—Key West Fragrance &
Cosmetic Factory, Inc., 1
Anti-Perspirant—Finelle Cosmetics, 2
D'Aloe Solid Deodorant—Key West Fragrance &
Cosmetic Factory, Inc., 1
Deodorant Cream—Shaklee U.S., Inc., 1
Deodorant—Gruene, 1
Desert Wind Deodorant—Shaklee U.S., Inc., 1
The Body Shop Deodorant—Body Shop, 2

Specialty Stores

Men's Range Stick Deodorant—Crabtree & Evelyn,
Ltd., 1

ANTISTATIC PRODUCTS

General Retail Stores

Shock Free—Magic American Corp., 1

ARTIFICIAL FINGERNAIL GLUE

General Retail Stores

Anti-Fungal Stick It Glue—Super Nail, 2
Instant Glue Remover—Ardell International, 2
Nails Off—Andrea International Ind., 2
Quick Stick Nail Glue—Ardell International, 2
Stick It Nail Glue—Super Nail, 2
Super Nails Liquids—Super Nail, 2
Super Nails Odorless Liquids and Powders—Super
 Nail, 2

Hair Salons

Lazer-Bond Primer—Alpha 9, 1
Nail, Tip & Glue Remover—Alpha 9, 1
Nail Powder—Alpha 9, 1
Odorless Liquid—Alpha 9, 1
10 Second Moveable Nail Adhesive—Alpha 9, 1

ARTIFICIAL FINGERNAILS

General Retail Stores

Andrea Nail Treatments—Andrea International
 Ind., 2

Classic Nails—Cosmyl Cosmetics, 3
Elektra Nails—Super Nail, 2
Exotic Nails—Cosmyl Cosmetics, 3
Nail Salon Sculptured Nail Kits—Ardell
    International, 2
Nail Tip Kits—Ardell International, 2
Nail-Wave—Super Nail, 2
Professional Nails—Andrea International Ind., 2
Sculptured Nail System—Super Nail, 2
Stick-on Nails—Ardell International, 2
Super Nails Tips and Kits—Super Nail, 2
Ultimate Acrylic Nail Kit—Super Nail, 2

ASTRINGENTS

General Retail Stores

Naturessence Water Lily Pore Astringent—Avanza, 1

Health Food Stores

Almond Astringent—Reviva Labs, 2
Citrus Essence Astringent—Kiss My Face, 1
Clarifying Herbal Astringent—Earth Science, 1
Eucalyptus Refining Astringent—Orjene Natural
    Cosmetics, 1
Herbal Facial Astringent—Aubrey Organics, 1
Lemon Mint Astringent—Rachel Perry, 1
Nature's Water Lily Pore Lotion Astringent—
    Avanza, 1
Peppermint Witch Hazel Astringent—Paul
    Penders, 1

Home Shopping/Mail Order

Astringent Lotion with Primrose & Chamomile—
    Kimberly Sayer, 1

Herb and Sulfur Astringent—Michel Constantini
    Natural Cosmetics, 2
Lily Herbal Astringent—Lily of Colorado, 1
Mint Astringent—Magic of Aloe, 3
Non-Alcoholic Astringent—Michel Constantini
    Natural Cosmetics, 2
Rose Astringent Lotion—Jeanne Rose, 2

Specialty Stores —————————————————

Four O'Clock Astringent—Body Shop (Calif.), 3
Oil Control Astringent— Natural Skin Care, 3

AUTOMATIC DISHWASHER SOAPS AND DETERGENTS

General Retail Stores ——————————————

Kleer II Automatic Dishwashing Detergent—
    Mountain Fresh Products, 1
Kleer III Powdered Automatic Dish Detergent—
    Mountain Fresh Products, 1

Health Food Stores ————————————————

Automatic Dishwashing Liquid—Sierra Dawn, 1
Kleer II Automatic Dishwasher Gel—Lotus Light, 1

Home Shopping/Mail Order ————————————

Automatic Dishwasher Detergent—Allens
    Naturally, 1
Basic-D Automatic Dishwashing Concentrate—
    Shaklee U.S., Inc., 1
Clean and Green Automatic Dishwashing Liquid—
    Ecco Bella, 1
Newmatic Automatic Dishwasher Detergent—
    Neway, 1

Professional Automatic Dishwashing Powder—
Home Service Products, 1

AUTOMOBILE PRODUCTS

Home Shopping/Mail Order

*All Purpose Industrial Grease—*KSA Jojoba, 1
Automatic Transmission Oil Lubricant—
KSA Jojoba, 1
*Engine Oil Lubricant—*KSA Jojoba, 1
*High Performance 2 Cycle Racing Lubricant—*KSA
Jojoba, 1
*Standard Transmission & Differential 90 Weight
Gear Lubricant—*KSA Jojoba, 1

BABY AND CHILDREN'S HAIR CONDITIONERS

General Retail Stores

*Baby Hair Lotion—*A.I.I. Clubman, 1
*Baby Massage Tearless Conditioner—*Mountain
Fresh Products, 1
*Kids Conditioner—*Kids William & Clarissa, 1

Health Food Stores

Baby Conditioner—Mera Mattis Beach, 1

BABY AND CHILDREN'S SHAMPOOS

General Retail Stores

Baby Massage Tearless Shampoo—Mountain Fresh Products, 1
Baby Shampoo—A.I.I. Clubman, 1
Kids Shampoo— Kids William & Clarissa, 1
No Tears Shampoo—Smith & Vandiver, Inc., 1

Hair Salons

Baby Don't Cry Shampoo—John Paul Mitchell Systems, 1

Health Food Stores

Baby mild + kind Shampoo—Borlind of Germany, 1
Baby Shampoo—Mera Mattis Beach, 1
Baby Shampoo — Extra Mild—Autumn Harp, 1
Natural Baby Shampoo—Aubrey Organics, 1
Nature's Gate Rainwater Herbal Baby Shampoo— Levlad, 1

Home Shopping/Mail Order

Gently Yours—Granny's Old Fashioned Products, 1

Specialty Stores

No Tears Shampoo—Scarborough and Company, 1
Tom Kitten Shampoo—Crabtree & Evelyn, Ltd., 1

BABY AND CHILDREN'S SOAPS

General Retail Stores

Baby Massage Gelly Wash—Mountain Fresh
   Products, 1
Baby Soap—Abbaco, Inc., 1
Kids Bar Soap—Kids William & Clarissa, 1
Kids Bubble Bath—Kids William & Clarissa, 1
Kids Liquid Soap—Kids William & Clarissa, 1

Health Food Stores

Baby mild + kind Children's Bath—Borlind of
   Germany, 1
Baby mild + kind Cleansing Milk—Borlind of
   Germany, 1
Calendula Baby Soap—Weleda, Inc., 1
Natural Baby Bath Soap—Aubrey Organics, 1

Specialty Stores

Alice in Wonderland Soaps—Crabtree & Evelyn,
   Ltd., 1
Beatrix Potter Figurine Soaps—Crabtree & Evelyn,
   Ltd., 1
Jemima Puddleduck Soap—Crabtree & Evelyn,
   Ltd., 1
Jeremy Fisher Soap—Crabtree & Evelyn, Ltd., 1
Liquid Body Wash—Scarborough and Company, 1
Peter Rabbit Soap—Crabtree & Evelyn, Ltd., 1
Tailor of Gloucester Soap—Crabtree & Evelyn,
   Ltd., 1
Tom Kitten Cleansing Lotion—Crabtree & Evelyn,
   Ltd., 1
Tom Kitten Liquid Soap—Cratbree & Evelyn,
   Ltd., 1
Tom Kitten Soap—Crabtree & Evelyn, Ltd., 1

BABY LOTIONS AND OILS

General Retail Stores

Baby Massage Massage Gelly—Mountain Fresh Products, 1

Baby Massage Massage Oil—Mountain Fresh Products, 1

Baby Massage Non-Oily Massage Lotion—Mountain Fresh Products, 1

Baby Oil—A.I.I. Clubman, 2

Kids Light Oil—Kids William & Clarissa, 1

Kids Lotion—Kids William & Clarissa, 1

Health Food Stores

Baby Lotion—Mera Mattis Beach, 1

Baby mild + kind Care Cream—Borlind of Germany, 2

Baby mild + kind Protection Cream—Borlind of Germany, 2

Baby mild + kind Skin Balm—Borlind of Germany, 2

Baby Oil with Lavender & Calendula—Lakon Herbals, 1

Baby Oil—No Mineral Oil—Liberty Products Corp., 1

Baby Oil—Petroleum Free—Autumn Harp, 1

Baby's Best Baby Bottom Balm—Liberty Natural Products Corp., 2

Calendula Baby Cream—Weleda, Inc., 2

Calendula Baby Oil—Weleda, Inc., 1

Natural Baby Body Lotion—Aubrey Organics, 1

Pure and Natural Baby Oil—Country Comfort, 1

Pure and Natural Herbal Baby Cream—Country Comfort, 1

Specialty Stores

Baby Lotion—Scarborough and Company, 1

Tom Kitten Cream—Crabtree & Evelyn, Ltd., 1
Tom Kitten Oil—Crabtree & Evelyn, Ltd., 1

BABY POWDERS

General Retail Stores

Baby Powder—Adwe Laboratories, 1
Good Clean Fun Powder—Smith & Vandiver, Inc., 1
Kids Powder—Kids William & Clarissa, 1

Health Food Stores

Baby Powder—Weleda, Inc., 1
Baby Powder—Talc Free—Autumn Harp, 1
Pure and Natural Herbal Baby Powder—Country
   Comfort, 1

Specialty Stores

Non-Talc Powder—Scarborough and Company, 1
Rice Bran Powder—Crabtree & Evelyn, Ltd., 1

BABY PRODUCTS— MISCELLANEOUS

General Retail Stores

Children's Sun Block—Wysong Corp., 1
Kids Boys and Girls Cologne—Kids William &
   Clarissa, 1
Kids Lip Balm—Kids William & Clarissa, 1
Kids Sun Block—Kids William & Clarissa, 1

Health Food Stores ─────────────────────────

All Smiles Baby Teething Oil—Liberty Natural
   Products, 1
Beach Baby—Reviva Labs, 2

BAR SOAPS

General Retail Stores ─────────────────────

Ace Tennis Soap— Abbaco, Inc., 3
Aerobics Glycerine Soap—Abbaco, Inc., 1
Algoli Facial Bar—Nature de France, 3
Aloe & Lanolin Skin Conditioning Soap—Andrew
   Jergens, 3
Ambi Cocoa Butter Soap—Kiwi Brands, Inc., 3
Ambi Glycerin Bar—Kiwi Brands, Inc., 3
Apiana Soaps—Baudelaire, 3
Argile Blanche Facial Bar—Nature de France, 3
Argile Rose Facial Bar—Nature de France, 3
Argilet Facial Bar—Nature de France, 3
Argimel Facial Bar—Nature de France, 3
Beautiful Glycerin Soaps—TerraNova, 1
Big B Glycerine Soap—Surrey, Inc., 3
Botanicals Glycerine Soap—Smith & Vandiver,
   Inc., 1
Chef's Soap—Abbaco, Inc., 3
Citronella Repellent Soap—North Country Soap, 1
Clearly Natural Glycerine Soap—Clearly Natural, 1
Country Soaps—North Country Soap, 1
CVS Glycerine Soap—Surrey, Inc., 3
De Maris Bath Soap—Cosmyl Cosmetics, 3
De Maris Guest Soap—Cosmyl Cosmetics, 3
Desert Rain For Men—TerraNova, 1
Faith in Nature Soaps—Baudelaire, 1
Fine Toiletries Glycerine Soap—Smith & Vandiver,
   Inc., 1
Fleurs des Champs—Abbaco, Inc., 1

Formula 142 Fisherman Soap—North Country Soap, 1

French Complexion Soap— Abbaco, Inc., 1

French Laundry Soap—Abbaco, Inc., 3

Gardeners Soap—Abbaco, Inc., 3

Gentle Cleansing Bar—Clientele, 1

Gentle Touch Soap—Andrew Jergens, 3

Glycerine Soap—Abbaco, Inc., 1

Golfer's Soap—Abbaco, Inc., 3

Gregory Aromatic Soaps—Baudelaire, 1

Herbal Soaps—North Country Soap, 2

Hog Wash The People Soap—North Country Soap, 1

Jergens Clear Complexion Bar—Andrew Jergens, 3

Jergens Mild Soap—Andrew Jergens, 3

Liberty Bar One Wash—North Country Soap, 1

McClinton's Barilla Soaps—Baudelaire, 1

Natural Bar Soap—Surrey, Inc., 3

New Soap—Botanicus, 1

North Country Soaps—North Country Soap, 1

Osco Glycerine Soap—Surrey, Inc., 3

Phebo Soaps—Baudelaire, 3

Pure Pleasure Soap—Surrey, Inc., 3

Rapid River Pine Tar—North Country Soap, 1

Runners Soap—Abbaco, Inc., 3

Scand/Ethnic Soaps—North Country Soaps, 1

Signature Soaps—Abbaco, Inc., 3

Sinclair & Valentine Glycerine Soap—Smith & Vandiver, Inc., 1

Sirena Soap—Tropical Soap Co., 1

Sportsman Soap—Abbaco, Inc., 3

Taylor Turkish Bath Soaps—Baudelaire, 1

Traveler Olive Oil Soap—Abbaco, Inc., 1

Vegetable Oil W/Honey Soaps—Abbaco, Inc., 1

Vitamin E & Lanolin Skin Conditioning Soap—Andrew Jergens, 3

Vitamin E Complexion Bar—Colonial Dames Co., Ltd., 1

Walgreens Glycerine Soap—Surrey, Inc., 3

Woodbury Soap—Andrew Jergens, 3

Workout Man-Sized Glycerine Soap—Abbaco, Inc., 1

Hair Salons ─────────────────

Biokosma Cleopatra Soap—Dr. Grandel, Inc., 2
Biokosma Sandalwood Soap—Dr. Grandel, Inc., 1
Biokosma Seaweed Pumice Soap—Dr. Grandel, Inc., 1

Health Food Stores ─────────────

All Vegetable Oil Glycerine Creme Soaps—Sappo Hill, 1
Almond Glow Beauty Bar—Home Health Products, 2
Aromatherapy Bath Soaps—Aura Cacia, 1
Beauty Soap—Beauty Without Cruelty, Ltd., 1
Camomile with Aloe Soap—Reviva Labs, 3
Castile Soap—Heritage Store, 1
Castor Oil Soap—Heritage Store, 1
Chandrika Ayurvedic Soap—Auroma International, 1
Chandrika Ayurvedic Soap—Auromere, Inc., 1
Coconut Oil Soap—Heritage Store, 1
Dr. Bronner's Magic Soap—All-One-God-Faith, 1
E-Gem Skin Care Soap—Carlson Laboratories, 3
Fresh Seaweed Soap—Natural Solutions, 3
Garden Fresh Soap—Carlson Laboratories, 1
G'Day Eucalyptus Bath Bar—Aubrey Organics, 1
Glycerine Soap—No Common Scents, 1
Glycerine Soap/Jojoba—No Common Scents, 2
Honeysuckle Vegetal Soap—Aubrey Organics, 1
Iris Soap—Weleda, Inc., 1
Meal and Herbs Exfoliation Skin Bar—Aubrey Organics, 1
Moisture Bar—Reviva Labs, 1
Oatmeal Soap—Reviva Labs, 3
Olive & Aloe Bar Soap—Kiss My Face, 1
Olive Oil Soap—Heritage Store, 1

Palma Christi Cleansing Bar Soap—Home Health
    Products, 2
Palmi Christi Deodorant Bar Soap—Home Health
    Products, 2
Pine Tar Soap—Heritage Store, 1
Pure Olive Oil Bar Soap—Kiss My Face, 1
Rainbow Aloe-Oatmeal Bar—Rainbow Research
    Corp., 1
Rainbow Clay Cleansing Bar—Rainbow Research
    Corp., 1
Rainbow Golden Moisture Bar—Rainbow Research
    Corp., 1
Rosa Mosqueta Complexion & Body Soap—Aubrey
    Organics, 1
Rose Soap—Weleda, Inc., 1
Rosemary Soap—Weleda, Inc., 1
Savon de Beaute Beauty Bars—Orjene Natural
    Cosmetics, 1
Sea Minerals Clay Soap—Sea Minerals, 1
Sea Minerals Soap—Sea Minerals, 1
Seaweed Soap—Reviva Labs, 2
Soothing Chamomile Soap—Natural Solutions, 3
Vegetable Oil Soap—NatureWorks, Inc.,1
Vegetable Oil Soaps—Alexandra Avery, 1
Vitamin Enriched Soap—Natural Solutions, 3

Home Shopping/Mail Order

Aloe Beauty Bar—Magic of Aloe, 3
Black Coral Bath Bar—Key West Fragrance &
    Cosmetic Factory, Inc., 1
Cinnamon Hand Cleaner—Brookside Soap, 1
Extra Mild Herbal Soap—Brookside Soap, 1
Extra Mild Unscented Soap—Brookside Soap, 1
Face & Body Soaps—Indian Creek, 2
Fandango Beauty Bar—Key West Fragrance &
    Cosmetic Factory, Inc., 1
F.F. Complexion Bar—The Face Food Shoppe, 1

Frangipani Bath Bar—Key West Fragrance &
   Cosmetic Factory, Inc., 1
Fresh Lime Soap—Brookside Soap, 1
Glycerin Fruit Soaps—Body Shop, 2
Heavenly Soaps—Heavenly Soap Co., 2
Jojoba Glycerin Soap—KSA Jojoba, 1
L'Arome Echoes Soap—L'Arome USA, Inc., 1
Lavenos Hand Soap—Livos Plant Chemistry, 1
Lemon Grass & Lime Soap—Brookside Soap, 1
Meadow Blend Soap-Free Cleansing Bar—Shaklee
   U.S., Inc., 1
Oatmeal & Almond—Brookside Soap, 1
Oliv Soap—Shahin Soap Co., 1
Oo La La Beauty Bar—Key West Fragrance &
   Cosmetic Factory, Inc., 1
Rosemary & Lavender Soap—Brookside Soap, 1
Scented Soap Selections—La Costa Products
   International, 1
Sea Caress Glycerine Soap—Wachter's, 1
Shahin Olive Oil Soap—Shahin Soap Co., 1
Spearmint Soap—Brookside Soap, 1
Vegetable Oil Soaps—Body Shop, 2
White Ginger Bath Bar—Key West Fragrance &
   Cosmetic Factory, Inc., 1

Specialty Stores ─────────────────────────

Almond Meal Soap—Crabtree & Evelyn, Ltd., 1
Fragranced Soap—Scarborough and Company, 1
Glycerin Soap—Potions & Lotions, 3
Maile Oval Drum Soap—Crabtree & Evelyn, Ltd., 1
Maize Meal Soap—Crabtree & Evelyn, Ltd., 1
Nutty Bar—i Natural Skin Care and Cosmetics, 3
Oatmeal Soap—Crabtree & Evelyn, Ltd., 1
Pikae Soap—Crabtree & Evelyn, Ltd., 1
Tiare Tahiti Soap—Crabtree & Evelyn, Ltd., 1
White Ginger Soap—Crabtree & Evelyn, Ltd., 1

BATHROOM CLEANERS

General Retail Stores ─────────────

> *Earth Friendly Cream Polish*—Venus Laboratories, Inc., 1
> *Earth Friendly Pine Soap Scrub*—Venus Laboratories, Inc., 1
> *Earth Rite Tub and Tile Cleaner*—Magic American Corp., 1
> *Tile 'n Grout Magic*—Magic American Corp., 1

Health Food Stores ─────────────

> *Shower Scour*—Natural Bodycare, 1

BATTERIES

General Retail Stores ─────────────

> *Duracell Batteries*—Duracell, Inc., 1
> *Energizer Batteries*—Eveready Battery Company, 1
> *Eveready Batteries*—Eveready Battery Company, 1

BEE PRODUCTS

General Retail Stores ─────────────

> *Apiana Pollen Creme*—Baudelaire, 2
> *Apiana Royal Jelly Creme*—Baudelaire, 2
> *Royal Bee Cream*—Colonial Dames Co., Ltd., 2

Health Food Stores ─────────────

> *Bee Healthy Propolis*—Liberty Natural Products, 2

Bee Healthy Royal Jelly—Liberty Natural
 Products, 2
Pollen Plus—Beehive Botanicals, 2
Propolis Capsules—Beehive Botanicals, 2
Propolis Chewing Gum—Beehive Botanicals, 2
Propolis Granules—Beehive Botanicals, 2
Propolis Tincture—Beehive Botanicals, 2
Royal Jelly—Beehive Botanicals, 2

BLEACHES AND WHITENERS—LIQUID

General Retail Stores

Miracle White—Kiwi Brands, Inc., 3
White 'N' Bright Laundry Additive—Soap Factory, 1
Winter White Non-Chlorine Liquid Bleach—
 Mountain Fresh Products, 1

Home Shopping/Mail Order

Newbrite Clothing Bleach—Neway, 1
Non-Chlorine Bleach—Seventh Generation, 1

BLEACHES AND WHITENERS—POWDERED

General Retail Stores

Rit Color Remover—Best Foods, 1

Rit Fabric Whitener and Brighteners—Best Foods, 1
Winter White Powdered Bleach—Mountain Fresh
    Products, 1

Health Food Stores _____

Golden Lotus Non-Chlorine Bleach—Lotus Light, 1

Home Shopping/Mail Order _____

*Nature-Bright Concentrated All-Fabric Laundry
    Brightener*—Shaklee U.S., Inc., 1
Professional All-Fabric Bleach—Home Service
    Products, 1

BODY TREATMENTS

Hair Salons _____

Biokosma Cellulite Seaweed Intensive—Dr. Grandel,
    Inc., 1
Biokosma Sea Line for Cellulite—Dr. Grandel, Inc., 1

Health Food Stores _____

Aztec Secret Healing Clay—Heritage Store, 1

Home Shopping/Mail Order _____

Contouring Body Mud—La Costa Products
    International, 1
Contouring Gel—La Costa Products International, 1
Purifying Body Mud—La Costa Products
    International, 1

BUBBLE BATHS, OILS, AND BATH TREATMENTS

General Retail Stores

ActiBath Carbonated Bath Tablets—Andrew Jergens, 3

Aloe Vera Herbal Bath—Colonial Dames Co., Ltd., 1

Bath Sea Salt—Botanicus, 1

Bathe Beans—Botanicus, 1

Birch Leaves Foaming Bathsoak—Nature de France, 3

Botanicals Bubble Bath—Smith & Vandiver, Inc., 1

Chamomile Foaming Bathsoak—Nature de France, 3

China Rose Bathing Bubbles—TerraNova, 1

Collagen & E Luxury Bath—Colonial Dames Co., Ltd., 1

Faith In Nature Essential Foam Bath—Baudelaire, 1

Floral Foaming Bath—Colonial Dames Co., Ltd., 1

Foaming Bath Creme—Botanicus, 1

Foaming Bath Oil Powder—Botanicus, 1

Gentle Touch Bath Beads—Andrew Jergens, 3

Good Clean Fun Bubble Bath—Smith & Vandiver, Inc., 1

Gregory Aromatic Bath Oils—Baudelaire, 1

Hubner Bath Oils—Baudelaire, 1

Jacob Hooy Chamomile Bath Foams—Baudelaire, 1

Jacob Hooy Jasmine Bath Foam—Baudelaire, 1

Jacob Hooy Lily of the Valley Bath Foam—Baudelaire, 1

Jacob Hooy Rose Bath Foam—Baudelaire, 1

Jasmine Bathing Bubbles—TerraNova, 1

Jergens Aloe & Lanolin Bath Beads—Andrew Jergens, 3

Jergens Bubbling Bath Beads—Andrew Jergens, 3

Jojoba Herbal Bath—Colonial Dames Co., Ltd., 1

Mineral & Herb Bath—Colonial Dames Co., Ltd., 1

Nature Scents Bath Beads—Andrew Jergens, 3
Organic Sea Salt—Cosmyl Cosmetics, 3
Pink Pearl Milk Bath—Colonial Dames Co., Ltd., 2
Sinclair & Valentine Bathing Crystals—Smith &
    Vandiver, Inc., 1
Sinclair & Valentine Bubble Bath—Smith &
    Vandiver, Inc., 1
Smith & Vandiver Bubble Bath—Smith &
    Vandiver, Inc., 1
Spice Foaming Bath—Colonial Dames Co., Ltd., 1
Strawberry Bathing Bubbles—TerraNova, 1
White Pearl Milk Bath—Colonial Dames Co., Ltd., 2

Hair Salons

Biokosma Aromatic Baths—Dr. Grandel, Inc., 1
Biokosma Cleopatra Milk Bath—Dr. Grandel, Inc., 2
Biokosma Seaweed Bath for Cellulite—Dr. Grandel,
    Inc., 1
Creamy Line Bath Bubbles—House of Lowell, Inc., 3
Lime Drops Bath Oil—House of Lowell, Inc., 3
Super Naturals Aloe Vera Bubble Bath—North
    Country Naturals, 2

Health Food Stores

Ancient Secret Dead Sea Bath Salts—Lotus Light, 1
Arizona Naturals Aloe Milk Bath—Arizona Natural
    Resources, Inc., 2
*Arizona Naturals Aloe Vera Bath & Shower Gel
    Concentrate*—Arizona Natural Resources, Inc., 2
Arizona Naturals Bath Oil Concentrate—Arizona
    Natural Resources, Inc., 2
Arizona Naturals Intensive Body Scrub—Arizona
    Natural Resources, Inc., 2
Aroma Beads—Body Love Natural Cosmetics, 3
Aromatherapy Baths—Aroma Vera, 1
Aromatherapy Mineral Baths—Aura Cacia, 1
Bath Moods Invigorating Foaming Bath Oil—
    Levlad, 1

Bath Moods Moisturizing Foaming Bath Oil—
    Levlad, 1
Bath Moods Relaxing Foaming Bath Oil—
    Levlad, 1
Bath Oils—Aroma Vera, 1
Bath Oils—Liberty Natural Products, 1
Body Satin Bath Oil—Earth Science, 1
Bubblegum Bubble Bath—Abracadabra, 1
Calming Flower Bath Oil—Paul Penders, 1
Camomile Bubbles Herbal Bath Oil—Aubrey
    Organics, 1
Cleopatra's Herbal Baths—WiseWays Herbals, 1
Concentrated Bath Oil—Jason Natural Products, 1
Dead Sea Mineral Bath—Sea Minerals, 1
Detox Bath—WiseWays Herbals, 1
Dragonberry Bubble Bath—Abracadabra, 1
Eucalyptus Spa Bath—Aubrey Organics, 1
French Lavender Body Bath—Alba Botanica, 1
Gorilla Bubble Bath—Abracadabra, 1
Hawaiian Seaweed Moisturizing Bath—Reviva
    Labs, 3
Herbomineral Ayurvedic Mud Bath Treatment—
    Auromere, Inc., 1
Honey Mango Body Bath—Alba Botanica, 2
Hula Bubble Bath—Abracadabra, 1
Keoki Papaya Bubble Bath—Natural Solutions, 3
Lavender Bath—Dr. Hauschka Cosmetics, 1
Lavender Bath Oil—Weleda, Inc., 1
Lemon Bath—Dr. Hauschka Cosmetics, 1
Love Mitts—Body Love Natural Cosmetics, 1
Luxury Bubble Baths—Abracadabra, 1
Mineral Bath—Abracadabra, 1
Peaceful Waters—WiseWays Herbals, 1
Pine Bath—Dr. Hauschka Cosmetics, 1
Pine Bath Oil—Weleda, Inc., 1
Pine Forest Bath Oil—WiseWays Herbals, 1
Pixieblossom Bubble Bath—Abracadabra, 1
Rainbow Bubble Bath—Rainbow Research Corp., 1
Relax-R-Bath Herbal Bath Emulsion—Aubrey
    Organics, 1

Rosa Mosqueta Bath Jalea—Aubrey Organics, 1
Rosemary Bath—Dr. Hauschka Cosmetics, 1
Rosemary Bath Oil—Weleda, Inc., 1
Sea Spa Bath Liquid—Aubrey Organics, 1
Sparkling Mint Body Bath—Alba Botanica, 1
Unicorn Bubble Bath—Abracadabra, 1

Home Shopping/Mail Order

Aloe Bubble Bath—Magic of Aloe, 3
Aromatherapy Bath Salts—The Face Food Shoppe, 1
Bath & Body Oils—Jeanne Rose, 2
Bath & Body Oils—Simplers Botanical Co., 1
Bath & Shower Foam Aromatherapy—O'Naturel, 1
Bath and Shower Gel—Michel Constantini Natural
    Cosmetics, 2
Bath/Body Oil—Heavenly Soap, 1
Bath Essence—Shaklee U.S., Inc., 1
Bath Formula Gelee—Irma Shorell, 2
Bath Herbal Tea Blends—Lady of the Lake
    Company, 1
Bath Herbs—The Face Food Shoppe, 1
Bath Milk—Shirley Price Aromatherapy, 1
Bath Oil—Body Shop, 2
Bath Oil Powder—Body Shop, 2
Bath Salts—Body Shop, 2
Bath Salts—Michel Constantini Natural
    Cosmetics, 2
Face Food Bath/Massage Oils—The Face Food
    Shoppe, 1
Herbal Baths—Jeanne Rose, 2
Honey Bath—The Face Food Shoppe, 2
Invigorating Bath and Shower Gel—Michel
    Constantini Natural Cosmetics, 2
Luxury Bath Oil Pearls—Oriflame International, 3
Milk Bath—Body Shop, 2
Orange Cream Bath Oil—Body Shop, 2
Raspberry Ripple Bathing Bubbles—Body Shop, 2
Roman Oil Beauty Bath—Patricia Allison, 1

Sea Bath Body Revitalizer—Wachter's, 1
Silky Bath Oil—Shirley Price Aromatherapy, 1
Skin-So-Soft—Avon Products, Inc., 3
Sun Milk—Shirley Price Aromatherapy, 1
Therapeutic Mineral Bath Salts—Michel
    Constantini Natural Cosmetics, 2
Tranquil Moments Aromatherapy—Avon Products,
    Inc., 3
Treatment Mixes—The Face Food Shoppe, 1

Specialty Stores

Bath Crystals—Potions & Lotions, 3
Bath Cubes—Crabtree & Evelyn, Ltd., 1
Bath Cubes—Scarborough and Company, 1
Bath Oils—Crabtree & Evelyn, Ltd., 1
Bath Pearls—Potions & Lotions, 3
Bath Powder—Potions & Lotions, 3
Bath Seeds—Crabtree & Evelyn, Ltd., 1
Boxed Bath Pearls—Scarborough and Company, 1
Bubble Bath/Shower Gel—Potions & Lotions, 3
China Rain Bubble Bath—Body Shop (Calif.), 3
Cocoa Butter Bath Oil—Potions & Lotions, 3
Cocoa Butter Bath Oil—Body Shop (Calif.), 3
Colored Bubble Bath—Body Shop (Calif.), 3
Fragranced Soap Leaves—Scarborough and
    Company, 1
Herbal Bath Totes—Scarborough and Company, 1
Milk Bath—Scarborough and Company, 2
Rain Bubble Bath—Body Shop (Calif.), 3
Spectrum Bubble Bath—Potions & Lotions, 3
Unscented Bubble Bath—Body Shop (Calif.), 3

C

CANDLES

General Retail Stores _____

Candle Gems—Botanicus, 1

Specialty Stores _____

Scented Candles—Scarborough and Company, 1

CARPET AND RUG DEODORIZERS

Home Shopping/Mail Order _____

Carpet Deodorizer—Seventh Generation, 1
Carpet Stuff—Rathdowney, Ltd., 1

CARPET AND RUG SHAMPOOS

General Retail Stores _____

Earth Friendly Rug Shampoo—Venus Laboratories, Inc., 1

Home Shopping/Mail Order _____

Carpet Shampoo—AFM Enterprises, Inc., 1
Karpet Kleen—Granny's Old Fashioned Products, 1

CARPET AND RUG SPOT AND STAIN REMOVERS

General Retail Stores _____

Carpet Stain Remover—Magic American Corp., 1
Just 'N Time Spot Remover—SerVaas
Laboratories, 1
Rug Stain and Spot Remover—Soap Factory, 1

CARPET PROTECTORS

Home Shopping/Mail Order _____

Carpet Guard—AFM Enterprises, Inc., 1

CAULK, SPACKLING, AND GROUT

Home Shopping/Mail Order _____

Caulking Compound—AFM Enterprises, Inc., 1
Joint Compound—AFM Enterprises, Inc., 1
Linseed-Putty—Livos Plant Chemistry, 1

Spackling Compound—AFM Enterprises, Inc., 1
Tile Grout—AFM Enterprises, Inc., 1
Vedo-Spackling Compound—Livos Plant
   Chemistry, 1

CHEST RUBS AND VAPORS

Health Food Stores —————————————

Chest Rub—NatureWorks, Inc., 2
Mama's Chest Rub—WiseWays Herbals, 2
Vapo-aid—Michael's Health Products, 1
Warm Up Rub—Breezy Balms, 2

Home Shopping/Mail Order —————————

Instant Medicated Vapor—Golden Pride/Rawleigh, 1

CLEANSERS

Home Shopping/Mail Order —————————

At-Ease Heavy-Duty Scouring Cleanser—Shaklee
   U.S., Inc., 1

COCOA BUTTER PRODUCTS

Health Food Stores —————————————

Cocoa Butter Creme w/Vitamin E—Jason Natural
   Products, 1

Light Touch/Light Mountain Cocoa Butter—Lotus
 Light, 1
100% Pure Cocoa Butter Super Moisturizer—Jason
 Natural Products, 1
Pure Cocoa Butter—Aura Cacia, Inc., 1

Specialty Stores _____

Cocoa Butter Cream—Body Shop (Calif.), 1
Pure Cocoa Butter—Body Shop (Calif.), 1

COFFEE FILTERS

General Retail Stores _____

Cone Coffee Filters—Melitta, 1
Melitta Natural Brown Basket Coffee Filters—
 Melitta, 1

COLD, FLU, AND HAYFEVER PRODUCTS

Health Food Stores _____

Aromatherapy—Heritage Store, 1
Campho-Derm—Heritage Store, 3
Flu and Cold Times—NatureWorks, Inc., 1
Inspirol—Heritage Store, 1
Koala Eucalyptus Inhalers—Liberty Natural
 Products, 1
Nose Balm—NatureWorks, Inc., 2
Ragweed Tincture—Heritage Store, 1

Similasan Drops for Colds & Influenza—Similasan
Corp., 2
Similasan for Colds—Similasan Corp., 1
Similasan Hayfever Drops No. 1—Similasan Corp., 1

Home Shopping/Mail Order

Cold Tablets—Golden Pride/Rawleigh, 1
Mouth & Throat Mist—Simplers Botanical Co., 1
Sinus Oil—Simplers Botanical Co., 1

COLOGNES AND PERFUMES

General Retail Stores

All colognes and fragrances—Clinique, Inc., 3
All colognes and fragrances—Estee Lauder, Inc., 3
Blue Waltz Perfume—A.I.I. Clubman, 1
Desert Rain Cologne for Men—TerraNova, 3
Eau de Parfum Sprays—TerraNova, 3
Jordi Eau de Toilette—Cosmyl Cosmetics, 3
Naim Eau de Cologne—Adwe Laboratories, 1
Perfume Essences—TerraNova, 1
Perfume Oil—Botanicus, 1
Tra La Cologne—Colonial Dames Co., Ltd., 1
Yafe Eau de Cologne—Adwe Laboratories, 1

Health Food Stores

Angelica Eau de Cologne—Aubrey Organics, 1
Arousing—Alexandra Avery, 1
Eau de Cologne—Weleda, Inc., 1
Elysian Fields Eau de Cologne—Aubrey Organics, 1
Glowing Touch Essential Perfume Oils—Sunshine
Products Group, 1
Grace—Alexandra Avery, 1
Ida Grae Earth Fragrance—Nature's Colors, 2

Influence—Alexandra Avery, 1
Je Suis Nuit Perfume—Earth Science, 1
Je Suis Perfume—Earth Science, 1
Lemon Blossom Body Splash Cologne—Aubrey
   Organics, 1
Musk Splash Eau de Cologne—Aubrey Organics, 1
Perfume Blends—No Common Scents, 1
Sunshine Essential Perfume Oils—Sunshine
   Products Group, 1
The Joyous—Alexandra Avery, 1
True Botanical Perfume Oils—Aura Cacia, 1
Wild Wind Eau de Cologne—Aubrey Organics, 1

Home Shopping/Mail Order

Bint El Sudan—Touchstone Company, 1
Black Coral Cologne—Key West Fragrance &
   Cosmetic Factory, Inc., 1
Broadway Baby—Key West Fragrance & Cosmetic
   Factory, Inc., 1
Classic Elegance Perfume—Key West Fragrance &
   Cosmetic Factory, Inc., 1
Classic Extravagance Perfume—Key West
   Fragrance & Cosmetic Factory, Inc., 1
Classic Femininity Perfume—Key West Fragrance
   & Cosmetic Factory, Inc., 1
Classic Plaisir Perfume—Key West Fragrance &
   Cosmetic Factory, Inc., 1
Classic Romance Perfume—Key West Fragrance &
   Cosmetic Factory, Inc.,
Classic Single Fragrances—Key West Fragrance &
   Cosmetic Factory, Inc., 1
Creme Perfumes—Patricia Allison, 1
Essentia One—Shirley Price Aromatherapy, 1
Everafter—Avon Products, Inc., 3
Fandango Cologne—Key West Fragrance & Cosmetic
   Factory, Inc., 1
Forever Eau de Parfum—Shaklee U.S., Inc., 1
Frangipani Cologne—Key West Fragrance &
   Cosmetic Factory, Inc., 1

Hibiscus & White Musk Oil Perfume—Key West Fragrance & Cosmetic Factory, Inc., 1

Hibiscus Cologne—Key West Fragrance & Cosmetic Factory, Inc., 1

Imari—Avon Products, Inc., 3

Jasmine Perfume—KSA Jojoba, 1

Key West Krazy—Key West Fragrance & Cosmetic Factory, Inc., 1

Lapis Cologne Spray—Shaklee U.S., Inc., 1

L'Arome Echoes Parfums—L'Arome USA, Inc., 1

Mary Kay Fragrances—Mary Kay Cosmetics, Inc., 3

Musk Perfume—KSA Jojoba, 1

Orange Perfume—KSA Jojoba, 1

Oo La La Cologne—Key West Fragrance & Cosmetic Factory, Inc., 1

Perfume Oil Concentrates—Patricia Allison, 1

Perfume Oils—Body Shop, 2

Replica's—Key West Fragrance & Cosmetic Factory, Inc., 1

Shaklee Whispers Cologne Spray—Shaklee U.S., Inc., 1

Soft Musk—Avon Products, Inc., 3

Tal Yen Cologne—Key West Fragrance & Cosmetic Factory, Inc., 1

Tea Rose—Body Shop, 2

Undeniable—Billy Dee Williams and Avon Products, Inc., 3

Vanilla Cologne—Key West Fragrance & Cosmetic Factory, Inc., 1

White Ginger Cologne—Key West Fragrance & Cosmetic Factory, Inc., 1

White Musk—Body Shop, 2

Specialty Stores

Colors—Benetton Cosmetics, 1

Designer Reproductions—Potions & Lotions, 3

Eau de Parfum—Scarborough and Company, 1

Floral Toilet Waters—Crabtree & Evelyn, Ltd., 1

Meridian Cologne—Scarborough and Company, 1

Perfumed Body Spray—Scarborough and
Company, 1

COSMETICS

General Retail Stores

All cosmetics—Adwe Laboratories, 1
All cosmetics—Clientele, 1
All cosmetics—Clinique Laboratories, Inc., 3
All cosmetics—Color Me Beautiful, 3
All cosmetics—Cosmyl Cosmetics, 3
All cosmetics—Estee Lauder, 3
All cosmetics—Natural Touch Cosmetics, 3
All cosmetics—Origins Natural Resources, Inc., 1
All cosmetics—Prescriptives, Inc., 3
IsaDora Cosmetique—Cosmetics Technology Inc., 2
Naturessence Color Cosmetics—Avanza Corp., 1

Hair Salons

All cosmetics—Studio Magic Cosmetics, 2
Indra Natural Colour Cosmetics—Aveda Corp., 1
Miliz Illusions Make-Up—Focus 21, 1
Il-Makiage Cosmetics—Il-Makiage, 3

Health Food Stores

All cosmetics—Aubrey Organics, 2
All cosmetics—Beauty Without Cruelty, Ltd., 1
All cosmetics—Borlind of Germany, 2
All cosmetics—Eva Jon Cosmetics, 3
All cosmetics—Orjene Natural Cosmetics, 3
All cosmetics—Rachel Perry, 3
Common Sense Cosmetics—Lotus Light, 1
Day Cosmetics—Dr. Hauschka Cosmetics, 2
Ida Grae Cosmetics—Nature's Colors, 1
Nature Cosmetics—Avanza Corp., 1
Water Base Color Cosmetics—Paul Penders, 1

Home Shopping/Mail Order

All cosmetics—Body Shop, 2
All cosmetics—Ecco Bella Cosmetics, 1
All cosmetics—Finelle Cosmetics, 2
All cosmetics—Magic of Aloe, 3
All cosmetics—Mary Kay Cosmetics, 3
All cosmetics—Michel Constantini Natural
   Cosmetics, 3
All cosmetics—Sombra Cosmetics, 1
All cosmetics—Warm Earth Cosmetics, 1
Avon Cosmetics—Avon Products, Inc., 3
Barbizon Cosmetics—Barbizon International, 3
Grace Cosmetics—Grace Cosmetics/Pro-Ma
   Systems, 1
Kallima Exquisite Glamour—Kallima
   International, Inc., 1
Shaklee Cosmetics—Shaklee U.S., Inc., 1

Specialty Stores

Bare Escentuals—Bare Escentuals/Dolphin
   Acquisition Corp., 2
Beaute Cosmetics—Benetton Cosmetics, 3

COUGH DROPS AND SYRUPS

Health Food Stores

Cough Syrup—NatureWorks, Inc., 1
Mother Earth's Cough Syrup—Heritage Store, 1
Similasan for Coughs—Similasan Corp., 1

CUTICLE CARE PRODUCTS

General Retail Stores

Andrea Cuticle Care—Andrea International Ind., 2

Ardell Cuticle Softener and Remover—Ardell International, 2
Buffing Cream—Super Nail, 2
Cuticare Cuticle Creme—Cosmyl Cosmetics, 3
Cuticare Cuticle Remover—Cosmyl Cosmetics, 3
Cuticle Conditioning Pen—Ardell International, 2
Cuticle Cream—Ardell International, 2
Cuticle Oil—Super Nail, 2
Cuticle Softener and Remover—Super Nail, 2
Cuticle Treat—DeLore Nails, 2
Cuticle Trimmer—Ardell International, 2
Cuticle Velvet Cream—Super Nail, 2
Nite Nail Cream—Super Nail, 2

Health Food Stores —————————————————

Nail Balm—Borlind of Germany, 1

Specialty Stores —————————————————

Cuticle Pusher—Crabtree & Evelyn, Ltd., 1

D

DANDRUFF SHAMPOOS AND CONDITIONERS

Hair Salons —————————————————

Paul Mitchell Tea Tree Special Shampoo—John Paul Mitchell Systems, 1

Top Tone Dandruff Shampoo—House of Lowell, Inc., 3

Health Food Stores

Aussie Gold Naturally Medicated Dandruff Conditioner—Jason Natural Products, 1
Aussie Gold Naturally Medicated Dandruff Shampoo—Jason Natural Products, 1
Everclean Dandruff Shampoo—Home Health Products, 2
Nature Therapeutic w/Antiseptic Tea Tree Oil Shampoo—Avanza Corp., 1
Nature's Gate Herbal Tea Tree Oil Shampoo—Levlad, 1
Nature's Therapeutic w/Antiseptic Tea Tree Oil Conditioner—Avanza Corp., 1
Seide Dandruff Shampoo—Borlind of Germany, 1
Tannebaum Pine Tar Shampoo—Heritage Store, 1

Home Shopping/Mail Order

Aloedan Dandruff Control Conditioner—Key West Fragrance & Cosmetic Factory, Inc., 1
Aloedan—Key West Fragrance & Cosmetic Factory, Inc., 1
Scalp & Dandruff Shampoo—Jeanne Rose, 2
Shampoo for Dandruff Hair—Michel Constantini Natural Cosmetics, 2

DELICATE FABRIC WASHES

General Retail Stores

Cool Wash—Mountain Fresh Products, 1
Delicate Laundry Detergent—Soap Factory, 1
Forever New—Forever New, 1

Hair Salons _____

Jurliques Gentle—Jurlique D'Namis, Ltd., 1

Home Shopping/Mail Order _____

Wool Wash Liquid—Ecover, 2

Specialty Stores _____

Cold Water Wash—Scarborough and Company, 3

DENTAL FLOSS

General Retail Stores _____

Tom's of Maine Natural Flossing Ribbon—Tom's of
  Maine, 1

Health Food Stores _____

Tea Tree Oil Dental Floss—Desert Essence, 1
Tea Tree Oil Dental Pics—Desert Essence, 1

DIET PRODUCTS

General Retail Stores _____

Hi-Fiber Wafers—Clientele, 1
Swiss Chocolaits—Clientele, 1
Weight Management Nutrients—Clientele, 1

Home Shopping/Mail Order _____

Slim Tea—Hobe Laboratories, Inc., 1

Swedish Supreme Fruit & Fiber Bar—Cernitin
America, Inc., 1
Swedish Supreme Nutritional Diet Mix—Cernitin
America, Inc., 1

DISHWASHING SOAPS

General Retail Stores

Dishwashing Liquid—Soap Factory, 1
Instant Dishwashing Concentrate—Adwe
Laboratories, 1
Kleer Dishwashing Detergent—Mountain Fresh
Products, 1

Health Food Stores

Golden Lotus Dishwashing Liquid—Lotus Light, 1
Kleer Dish Soap—Lotus Light, 1
Liquid Dish Soap—Ecover, 1
New America Dishwashing Liquid—Abkit, Inc., 1
Premium Dishwashing Liquid—Sierra Dawn, 1

Home Shopping/Mail Order

Aloe Care Dish/All Purpose Liquid—Granny's Old
Fashioned Products, 1
Clean and Green Dishwashing Liquid—Ecco
Bella, 1
Dishwashing Liquid—Allens Naturally, 1
Dishwashing Liquid—Seventh Generation, 1
New Liquid Suds Dish Washing Detergent—
Neway, 1
Professional Dishwashing Liquid—Home Service
Products, 1
Satin Sheen Dishwashing Liquid—Shaklee U.S.,
Inc., 1

DRAIN OPENERS AND CLEANERS

General Retail Stores ─────────────

Earth Friendly Grease Control—Venus
  Laboratories, Inc., 1
Natumate Enzymatic Drain Opener—Venus
  Laboratories, Inc., 1

Home Shopping/Mail Order ─────────────

Bio-Free Drain Opener—Seventh Generation, 1
Liquid Enzyme Drain Cleaner—Ecco Bella, 1

E

EMERY BOARDS AND BUFFING TOOLS

General Retail Stores ─────────────

Ardell Emery Disks—Ardell International, 2
Ardell Emery Hearts—Ardell International, 2
Natural Chamois Buffer—Ardell International, 3
Quick Shine Buffers—Ardell International, 2
Space-eez Finger/Toe Spacers—Ardell
  International, 2
Super Nails Emery Boards and Files—Supor
  Nails, 2

Specialty Stores _____

Emery Boards—Crabtree & Evelyn, Ltd., 1

ESSENTIAL OILS

Hair Salons _____

Essential Oils—Jelene International, 1

Health Food Stores _____

Amber Essence—Eden Botanicals, 1
Aromatherapy Escentual Oil—Bare Escentuals/
    Dolphin Acquisition Corp., 2
Aromatic Essential Oils—Eden Botanicals, 1
Auroshikha Essential Oils—Auroma
    International, 1
Blends of Essential Oils—Aroma Vera, 1
Essential Oils—Aroma Vera, 1
Essential Oils—Aura Cacia, 1
Essential Oils—Liberty Natural Products, 1
Essential Oils—No Common Scents, 1
Herbal Oils—WiseWays Herbals, 1
Nature's Alchemy Essential Oils—Lotus Light, 1
100% Pure Essential Oil Aromatherapy Blends—
    Aura Cacia, 1
Organic Herbal Tincture and Oils—Eden
    Botanicals, 1
Rare & Precious Oils—Aroma Vera, 1
Sunshine Herbal Comfort Oils—Sunshine Products
    Group, 1
Water-Soluble Aromatics—Aroma Vera, 1

Home Shopping/Mail Order _____

Annointing Oils—The Face Food Shoppe, 1

Aromatherapy Blends—Lady of the Lake
Company, 1
Aromatherapy Oils—The Face Food Shoppe, 1
Compound Glycerites—Simplers Botanical Co., 1
Essential & Perfume Oils—Michel Constantini
Natural Cosmetics, 2
Fixed & Essential Oil Compounds—Simplers
Botanical Co., 1
MLE Essential Essence—Wachter's, 1
Pure Essential Oils—Shirley Price Aromatherapy, 1
Pure Oil of Kukui—Lily of Colorado, 1
Seven Exotic Oils—Lily of Colorado, 1

Specialty Stores

Enviromental Oils with Dropper—Scarborough and
Company, 1
Escentual Perfume Oils—Bare Escentuals/Dolphin
Acquisition Corp., 2
Essences—Potions & Lotions, 3
Special Essences—Potions & Lotions, 3

EXFOLIATING SCRUBS AND PEELS

General Retail Stores

Exfoliating Scrub—Color Me Beautiful, 1
Gentle Sloughing Creme—Color Me Beautiful, 2
Naturessence Non Chemical Skin Peel—Avanza
Corp., 1
Vital Exfoliating Scrub—Cosmyl Cosmetics, 3

Hair Salons

L'Exfoliant—Mastey de Paris, 1

Health Food Stores

Aloegen Papaya Scrub Exfoliante—Levlad, 3
Apricot Kernel Oil Facial Scrub—Orjene Natural
Cosmetics, 2
Bio-Peeling Masque—Borlind of Germany, 2
Eva Jon Facial Scrub—Eva Jon Cosmetics, 3
Gentle Body Smoother—Alba Botanica, 1
Geremy Rose Avena Oat Scrub—New Moon
Extracts, Inc.,
Honey & Almond Scrub—Orjene Natural
Cosmetics, 2
Kilauea Volcanic Pumice Scrub—Natural
Solutions, 3
Manteca Almond Honey Body Scrub—Natural
Solutions, 3
Nature Non Chemical Peel—Avanza Corp., 1

Home Shopping/Mail Order

Aloe in the Pink—Key West Fragrance & Cosmetic
Factory, Inc., 1
Body Polishing Scrub—La Costa Products
International, 1
Cleansing Grains—Michel Constantini Natural
Cosmetics, 2
Dermipeel—Magic of Aloe, 3
Facial Scrub—Key West Fragrance & Cosmetic
Factory, Inc., 1
Honey/Almond Facial Scrub—Magic of Aloe, 3
Honey & Almond Scrub—Jacklyn Cares, 1
Honey and Almond Scrub—Michel Constantini
Natural Cosmetics, 2
Honey & Almond Scrub—Nutri-Metrics
International, Inc., 1
Japanese Washing Grains—Body Shop, 2
Multi-Active Exfoliating Scrub—Kallima
International, Inc., 1

Therapeutic Buffing Cream—Michel Constantini
   Natural Cosmetics, 2

Specialty Stores

Camphor Blend Ex-Foligel—Potions & Lotions, 3
Exfoliating Body Scrub—Scarborough and
   Company, 1
Honey-Almond Scrub—Potions & Lotions, 3
Nuts-A Scrub—i Natural Skin Care and
   Cosmetics, 3
Oil Control Scrub Wash—i Natural Skin Care and
   Cosmetics, 3
Sweet Corn Body Sloughing Cream—i Natural Skin
   Care and Cosmetics, 3
Therapeutic Buffing Creme—Potions & Lotions, 3

EYE AND THROAT CREAMS

General Retail Stores

Eye Contour Cream for the Night—Jean Pax, Inc., 1
Eye Contour Gel—Jean Pax, Inc., 1
Fortifying Eye Cream—Color Me Beautiful, 3
Hydra-Cell Eye Creme—Cosmyl Cosmetics, 3
Naturessence Anti-Wrinkle Eye & Throat Cream—
   Avanza Corp., 1
Refreshing Eye Treat—Color Me Beautiful, 1

Hair Salons

Eye Creme—Mastey de Paris, 1
Super Naturals Elastin Eye Cream—North Country
   Naturals, 3

Health Food Stores

Azulene Desensitizing Eye Cream—Earth Science, 1

Blackmores Angelica Eye Creme—Solgar Vitamin Co., 1

Blackmores Cornflower Eye Balm—Solgar Vitamin Co., 1

Calming Chamomile Eye Gel—Natural Solutions, 3

Carotene Eye Gelee with A & E—Paul Penders, 1

Delicate Extra-Rich Eye Cream—Natural Bodycare, 1

Eye Beauty Cream—Orjene Natural Cosmetics, 3

Eye Freshener—Dr. Hauschka Cosmetics, 1

Eye Gelee Concentrate—Reviva Labs, 3

Eye Lid Cream—Dr. Hauschka Cosmetics, 2

Eye Oil—Zia Cosmetics, 3

Eye Wrinkle Cream—Borlind of Germany, 2

Eye-Q Eye Area Moisturizer—Jason Natural Products, 1

Gelle For Eyes—Orjene Natural Cosmetics, 3

Ida Grae Earth Eye/Lip Creme—Nature's Colors, 2

Irish Moss Eye Gel—Lily of Colorado, 1

LL Eye Wrinkle Cream—Borlind of Germany, 2

Nature Anti Wrinkle Eye & Throat Cream—Avanza Corp., 1

100% Collagen Fiber Eye Pads—Reviva Labs, 3

Placentagen Eye and Throat Cream—Earth Science, 1

Santa Barbara Avocado Eye Cream—Natural Solutions, 3

Throat & Eye Cream—Reviva Labs, 2

Vitamin E Oil Stick for Eyes/Lips—Reviva Labs, 2

Home Shopping/Mail Order

Aloe Eye Cream w/R.N.A.—Grace Cosmetics/ Pro-Ma Systems, 1

Avocado Eye Creme—Nutri-Metrics International, Inc., 1

Chickweed Eye Cooler—The Face Food Shoppe, 2

Elderflower Under Eye Gel—Body Shop, 2

Energizing Eye Gelee—Michel Constantini Natural Cosmetics, 2

Eye Calm Gel Soother—Oriflame International, 3
Eye Creme—Finelle Cosmetics, 2
Eye Creme Concentrate—Mary Kay Cosmetics, Inc., 3
Eye Creme Plus—Wachter's, 1
Eye Perfector—Avon Products, Inc., 3
Eye Renewal Cream with Collagen—Kimberly Sayer, Inc., 1
Eye Rescue—Finelle Cosmetics, 2
Eye-Line Creme—Wachter's, 1
EyeMoist—Key West Fragrance & Cosmetic Factory, Inc., 2
Fibronectin Eye Cream—Michel Constantini Natural Cosmetics, 3
Firming Eye Gel—La Costa Products International, 1
Hyaluronic Eye Creme—Michel Constantini Natural Cosmetics, 3
Moisturizing Eye Primer—Oriflame International, 3
Neck Gel—Body Shop, 2
Nourishing Eye Creme—La Costa Products International, 1
Protective Delicate Zone Moist Dewy—Kallima International, Inc., 1
Secret of Switzerland Throat/Eye Creme—Patricia Allison, 2
Super Eye Moisturizer—Magic of Aloe, 3
Superlight Eye Cream—Shirley Price Aromatherapy, 1

Specialty Stores

Eye Cream—Body Shop (Calif.), 3
Eye Profile Cream—i Natural Skin Care and Cosmetics, 3
Eye Profile Gel—i Natural Skin Care and Cosmetics, 3
Golden Eye Gel—Potions & Lotions, 3

EYE DROPS

Health Food Stores

Optikade Eye Tonic—Heritage Store, 1
Similasan for Eyes Drops No. 1—Similasan Corp., 1
Similasan for Eyes Drops No. 2—Similasan Corp., 2

Home Shopping/Mail Order

Chamomile Eye Drops—Shirley Price
Aromatherapy, 1

EYELASHES AND EYELASH PRODUCTS

General Retail Stores

Ardell Duralash—Ardell International, 2
Ardell Fashion Lashes—Ardell International, 2
Ardell Invisibands—Ardell International, 2
Ardell Lashlites—Ardell International, 2
Duralash Lashfree Remover—Ardell
International, 2
Duralash Lashlite Adhesive—Ardell International, 2
Lash Adhesive—Andrea International Ind., 2
Lashcare Cleaner—Ardell International, 2
Lashgrip Adhesive—Ardell International, 2
Mod Lashes—Andrea International Ind., 2
Mod Strip Lashes—Andrea International Ind., 2
Talika Eyelash Conditioning Cream—Jean Pax,
Inc., 1
Talika New Eyelash Conditioning Gel—Jean Pax,
Inc., 1

F

FABRIC SOFTENERS

General Retail Stores ———————————————

Soft-N-Fresh—Mountain Fresh Products, 1

Health Food Stores ———————————————

Fabric Conditoner Softener—Ecover, 1
Soft-N-Fresh—Lotus Light, 1

Home Shopping/Mail Order ———————————

Concentrated Fabric Softener—Allens Naturally, 1
Professional Pre-Add Fabric Softener—Home
Service Products, 1
Softer Than Soft Concentrated Fabric Conditioner—
Shaklee U.S., Inc., 1

FACIAL CLEANSERS

General Retail Stores ———————————————

All Purpose Cream—Colonial Dames Co., Ltd., 1
Ambi Complexion Soap—Kiwi Brands, Inc., 3
Ambi Skin Cleaning Creme—Kiwi Brands, Inc., 3

Apiana Honey Cleansing Milk—Baudelaire, 2
Botanical Cleansing Gel—Smith & Vandiver, Inc., 1
Cleanser I—Color Me Beautiful, 2
Deep Pore Cleansing Milk—Cosmyl Cosmetics, 3
Delicate Cleansing Milk—Cosmyl Cosmetics, 3
Dry Skin Cleansing Creme—Colonial Dames Co.,
    Ltd., 1
Dry Skin Cleansing Lotion—Colonial Dames Co.,
    Ltd., 1
Elastin & NaPCA Creamy Facial Wash—Avanza
    Corp., 1
Extra Gentle Cleansing Creme—Color Me
    Beautiful, 1
Face & Body Cleanser/Masque—La Crista, Inc., 1
Facial Cleansing Gel—Botanicus, 1
Honey & Almond Refining Scrub—TerraNova, 3
Honey & Almond Scrub—Tyra Skin Cleaner, 2
Mint Refining Scrub—TerraNova, 3
Natural Buff & Glycerine—Surrey, Inc., 3
Natural Touch Cleansing Bar—Natural Touch, 3
*Naturessence Antiseptic Clarifying Deep Cleansing
    Lotion*—Avanza Corp., 1
Naturessence Apricot & Kelp Facial Scrub—Avanza
    Corp., 1
*Naturessence Citrus and Aloe Vera Cleansing
    Milk*—Avanza Corp., 1
Naturessence Fresh Lemon Cleansing Cream—
    Avanza Corp., 1
Premiere Cleanser—Tyra Skin Care, 2
Skin Conditioning Cleanser—Color Me Beautiful, 1
Truly Moist Cleanser—Tyra Skin Care, 2
Very Effective Cleansing Gel—Color Me Beautiful, 2
Vitamin E Antiseptic Cleanser—Colonial Dames
    Co., Ltd., 1
Vitamin E Moisturizing Cleanser—Colonial Dames
    Co., Ltd., 1

Hair Salons _____

Cleansing Cream—Aveda Corp., 1

Cleansing Gel—Jelene International, 1
Cleansing Lotion—Jelene International, 1
Cleansing Scrub—Aveda Corp., 1
Detox21 Skin Cleanser—Focus 21, 1
Emulsione—Mastey de Paris, 1
Energizer Scrub—Jelene International, 1
Facial Clarifier—Aveda Corp., 1
Hy-Oil Cleanser—Dr. Babor Natural Cosmetics, 2
L'Stimuli—Mastey de Paris, 1
Mild Cleanser—Dr. Babor Natural Cosmetics, 1
Stuff Wash—Jelene International, 1
Super Naturals Aloe Vera Cleansing Cream—North
    Country Naturals, 2
Super Naturals Apricot Facial Scrub—North
    Country Naturals, 2
Super Naturals Elastin Cleansing Cream—North
    Country Naturals, 3
Super Naturals Vitamin E Cleansing Cream—North
    Country Naturals, 2

Health Food Stores

A-D-E Creamy Cleanser—Earth Science, 1
Aloe & Olive Cleansing Cream—Beauty Without
    Cruelty, Ltd., 1
Aloe Herb Facial Cleanser—Natural Bodycare, 1
Aloe Vera Perfect Cleansing Creme—Orjene Natural
    Cosmetics, 3
Aloe-Herb Facial Scrub—Natural Bodycare, 1
Aloegen Clarifying Cleansing Emulsion—Levlad, 3
Aloegen Revitalizing Body Scrub—Levlad, 3
Amazing Grains—Body Love Natural Cosmetics, 1
Apricot Mild Facial Scrub Cream—Earth Science, 1
Apricot Mild Facial Scrub Lotion—Earth Science, 1
Apricot Scrubble Facial Wash & Scrub—Jason
    Natural Products, 2
Aqualin Cleanser—Micro Balanced Products
    Corp., 1
Arizona Naturals Light Facial Scrub—Arizona
    Natural Resources, Inc., 2

Arizona Naturals Total Cleanser—Arizona Natural
Resources, Inc., 2

Blackmores Almond Cleansing Creme—Solgar
Vitamin Co., 1

Blackmores Cucumber Cleanser—Solgar Vitamin
Co., 1

Blackmores Marshmallow Complexion Soap—
Solgar Vitamin Co., 1

Blakemores Cinnamon Scrub—Solgar Vitamin
Co., 1

Calendula Comfrey Cleansing Milk—Paul
Penders, 1

Camphor Cleanser Lotion—Orjene Natural
Cosmetics, 2

Camphor Cleansing Milk—Reviva Labs, 2

Chamomile & Wheat Germ Oil Cleanser Lotion—
Orjene Natural Cosmetics, 1

Citrus-Aloe Cleanser & Face Wash—Rachel
Perry, 1

Clarifying Facial Wash—Earth Science, 1

Cleansing Milk—Reviva Labs, 2

Cleansing Cream—Dr. Hauschka Cosmetics, 1

Complexion Care—Alexandra Avery, 1

Creamy Facial Cleanser—Alba Botanica, 1

Cucumber Cleansing Milk—Beauty Without
Cruelty, Ltd., 1

Extra Gentle Face Scrub—Alba Botanica, 1

Facial Cleansing Bar—Beauty Without Cruelty,
Ltd., 1

Facial Cleansing Gel—Home Health Products, 2

Flowers of Lilac Cleansing Cream—Beauty Without
Cruelty, Ltd., 1

Foaming Herbal Face Wash—Beauty Without
Cruelty, Ltd., 1

Gentle Loofa Face Scrub—Beauty Without Cruelty,
Ltd., 1

Geremy Rose Creamy Rose Cleanser—New Moon
Extracts, Inc., 1

Geremy Rose Santa Ana Cleanser—New Moon
Extracts, Inc., 1

Ginseng Face Cream—Aubrey Organics, 1

Ginseng Face Scrub—Aubrey Organics, 1

Herbal Antiseptic Skin Cleanser—Rachel Perry, 1

Herbal Cleansing Milk—Borlind of Germany, 1

Herbal Face Wash—Beauty Without Cruelty, Ltd., 1

Herbal-Response Liquid Cleanser—Natural Bodycare, 1

Honey-Almond Scrub with Oat Flour & Bran—Reviva Labs, 2

Iris Cleansing Lotion—Weleda, Inc., 2

Jojoba Walnut Facial Scrub—Desert Essence, 1

Kentucky Lemongrass Cleanser—Natural Solutions, 1

Lemon and Sage Cleansing Emulsion—Beauty Without Cruelty, Ltd., 1

LL Cleansing Milk—Borlind of Germany, 1

Mandarin Orange Facial Cleanser—Aubrey Organics, 1

Natural Organic Facial Cleanser—Aubrey Organics, 1

Nature Apricot & Kelp Facial Scrub—Avanza Corp., 1

Nature Fresh Lemon Deep Pore Cleansing Creme—Avanza Corp., 1

Nature's Aloe Vera & NaPCA Creamy Facial Wash—Avanza Corp., 1

Nature's Antiseptic Clarifying Deep Cleansing Lotion—Avanza Corp., 1

Nature's Citrus & Aloe Vera Cleansing Milk—Avanza Corp., 1

Nature's Gate Facial Cleanser—Levlad, 1

Nature's Gate Facial Scrub—Levlad, 1

Olive & Aloe Cleansing Creme—Kiss My Face, 1

Peppermint Juniper Cleansing Gel—Paul Penders, 1

Really Clean NaPCA Facial Cleanser—Desert Naturels, 1

Rosa Mosqueta Moisturizing Cleansing Bar—Aubrey Organics, 1

Rose Dew Cleansing Milk—Borlind of Germany, 1

Rosemary Elderflower Cleansing Milk—Paul
    Penders, 1
Sea Kelp-Herbal Facial Scrub—Rachel Perry, 2
Sea Ware Facial Cleanser—Aubrey Organics, 1
Seasoap Face & Body Cleansing Cream—Aubrey
    Organics, 1
Sensitive Skin Cleansing Gel—Reviva Labs, 2
Serious Skin Daily Cleanser—Natural Solutions, 3
Silica Scrub—Reviva Labs, 2
Special Cleansing Cream—Orjene Natural
    Cosmetics, 2
Strawberry Cleansing Cream—Orjene Natural
    Cosmetics, 2
Super Rich Cleansing Emulsion—Reviva Labs, 2
Sweet Honey Cream Cleanser—Natural Solutions, 3
Ultra Mild Liquid Cleanser—Natural Bodycare, 1
Wild Passion Flower Facial Polish—Natural
    Bodycare, 1
Z Herbal Cleansing Milk—Borlind of Germany, 1
Z 1 Moisturizing Cleanser—Zia Cosmetics, 1
Z 2 Fresh Cleansing Gel—Zia Cosmetics, 1

Home Shopping/Mail Order

Active Cleanser—Oriflame International, 3
All-Type Skin Cleanser—Michel Constantini
    Natural Cosmetics, 2
Aloe & Panthenol Cleansing Cream—Sombra, 1
Aloe Clean—Nutri-Metrics International, Inc., 1
Aloe Cleansing Lotion #1—Grace Cosmetics/
    Pro-Ma Systems, 1
Aloe Deep Cleanser—Grace Cosmetics/Pro-Ma
    Systems, 1
Aloe Facial Stimulating Scrub—Grace Cosmetics/
    Pro-Ma Systems, 1
Aloe Milk Cleanser—Michel Constantini Natural
    Cosmetics, 2
Aloe Skin Cleanser—Magic of Aloe, 3
Aloe Skin Cleanser Lotion—Jacklyn Cares, 2

Citrus Cleansing Lotion—Key West Fragrance & Cosmetic Factory, Inc., 1

Clarifying Lotion—Wachter's, 1

Cleansing Butter—The Face Food Shoppe, 2

Cleansing Cream—Finelle Cosmetics, 2

Cleansing Cream—Shirley Price Aromatherapy, 1

Cleansing Cream—The Face Food Shoppe, 2

Cleansing Cream with Chamomile & Primrose—Kimberly Sayer, Inc., 1

Cleansing Creme—World of Aloe, 1

Cleansing Lotion—Bronson Pharmaceuticals, 1

Cleansing Milk—Shirley Price Aromatherapy, 1

Collagen Cleansing Cream—Magic of Aloe, 3

Colloidal Cleanser—Biogime, 1

Conditioning Cleanser—Shaklee U.S., Inc., 1

Creamy Cleanser—Mary Kay Cosmetics, Inc., 3

Cucumber Cleanser—Michel Constantini Natural Cosmetics, 2

Cucumber Cleansing Milk—Body Shop, 2

Deep Acting Facial Cleanser—Irma Shorell, 2

Deep Cleanser—Mary Kay Cosmetics Inc., 3

Deep Cleansing Cream—Kimberly Sayer, Inc., 1

Dermabrase/35—Irma Shorell, 2

Desert Mist Cleanser—Hobe Laboratories, Inc., 1

Extra Emollient Cleansing Cream—Mary Kay Cosmetics, Inc., 3

Facial Cleansing Bar—Shaklee U.S., Inc., 1

Facial Scrub—KSA Jojoba, 1

Foamy Herbal Cleanser—Michel Constantini Natural Cosmetics, 2

Formula for Cleansing—Irma Shorell, 2

French Strawberry Creme Cleanser—Patricia Allison, 1

Gentle Cleansing Cream—Mary Kay Cosmetics, 3

Gentle Cleansing Lotion—Shaklee U.S., Inc., 1

Glycerin & Oatmeal Lather—Body Shop, 2

Herbal Facial Steam Mixtures—Jeanne Rose, 2

Honey & Almond Scrub—World of Aloe, 2

Honey, Beeswax, Almond & Jojoba Oil Cleaner—
Body Shop, 2

Inner-Cellular Cleansing Emulsion—Kallima
International, Inc., 1

Inter-Active Herbal Facial Wash—Kallima
International, Inc., 1

Lathering Cleanser—Biogime, 1

Lily Facial Cleanser—Lily of Colorado, 1

Liqui-Facial Wash—Magic of Aloe, 3

Mint Cleanser—Magic of Aloe, 3

Moisan Clean Glow—Nutri-Metrics International,
Inc., 1

Morning Cleansing Refresher—Oriflame
International, 3

Natural Apricot Body Scrub—Wachter's, 1

Natural Apricot Facial Wash—Wachter's, 1

Oats Honey Facial Scrub—Sombra, 1

Oats-N-Erbs—The Face Food Shoppe, 1

Orange Blossom Cleansing Milk—G. T.
International, 1

Orange Cleanser—Michel Constantini Natural
Cosmetics, 2

Orchid Oil Cleansing Milk—Body Shop, 2

Panthenol Cleanser—Michel Constantini Natural
Cosmetics, 2

Papaya Revitalizing Scrub—Indian Creek, 1

Passion Fruit Cleansing Gel—Body Shop, 2

Pineapple Facial Wax—Body Shop, 2

Proteinized Cleanser—Shaklee U.S., Inc., 1

Purifying Bar—Mary Kay Cosmetics, Inc., 3

Revitalizing Cleanser—Wachter's, 1

Revitalizing Scrub—G. T. International, 1

Royal Cleansing Grains—Earth Gifts, 1

Skin Clean—Nutri-Metrics International, Inc., 1

Strawberry Facial Wash—Barbizon International, 3

Swedish Pollenique Deep Cleansing Milk—Cernitin
America, Inc., 1

Swedish Pollenique Essential Skin Wash—Cernitin
America, Inc., 1

Swedish Scrub—Patricia Allison, 2
Velva Sea Cleansing Creme—Wachter's, 1
Washing Cleanser/Damaged Skin—Magic of Aloe, 3

Specialty Stores

Azulene Cleansing Lotion—Potions & Lotions, 3
Clarifying Cleanser—Body Shop (Calif.), 1
Cleansing Lotion—Body Shop (Calif.), 3
Cream Wash—Body Shop (Calif.), 3
Creamy Cleanser—i Natural Skin Care and
    Cosmetics, 3
Cucumber Cleanser—i Natural Skin Care and
    Cosmetics, 1
Dermaloge Cleansing Lotion—Potions & Lotions, 3
Facial Meal Scrub—Scarborough and Company, 1
Foaming Lotion Cleanser—Potions & Lotions, 3
Lemon Oat Facial Cleanser—Body Shop (Calif.), 3
Oil Control Gel Wash—i Natural Skin Care and
    Cosmetics, 1
Oil Control Scrub Wash—i Natural Skin Care and
    Cosmetics, 3
Oil Free Facial Cleanser—Body Shop (Calif.), 3
Strawberry Fluff Cleanser—i Natural Skin Care
    and Cosmetics, 3

FACIAL MASKS

General Retail Stores

Andrea At Home Spa Facial Masque—Andrea
    International Ind., 2
China Clay Soothing Mask—TerraNova, 3
Clay & Mineral Purifying Mask—Avanza Corp., 1
Facial Masque—Clientele, 1
Green Clay Deep Cleansing Facial Masque—Nature
    de France, 3

Honey & Clay Purifying Masque—Cosmyl
    Cosmetics, 3
Hydra-Cell Wheat Germ Masque—Cosmyl
    Cosmetics, 3
Hydrating Mask—Color Me Beautiful, 3
Regulating Mask—Color Me Beautiful, 1

Hair Salons

Biokosma Activating Mask—Dr. Grandel, Inc., 1
Calmative Masque—Aveda, 1
Cream Peeling—Dr. Babor Natural Cosmetics, 1
Cremask—Mastey de Paris,1
Electro-Gel 21 Skin Activating Mask—Focus 21, 1
Facial Rejuvenating Masque Powder—Aveda
    Corp., 1
Iced Creme Facial Masque—Iced Creme Facial
    Masque, 2
Skin Vitalizer—Dr. Babor Natural Cosmetics, 1
Vitalizing Masque—Aveda Corp., 1

Health Food Stores

Almond Mask—Reviva Labs, 2
Aloegen Sea Kelp Firming Mask—Levlad, 3
Ancient Secret Mud from the Dead Sea—Lotus
    Light, 1
Blackmores Herbal Clay Face Masque—Solgar
    Vitamin Co., 1
Camphor Treatment Mask—Zia Cosmetics, 1
Clay & Ginseng Texturizing Mask—Rachel Perry, 1
Clay Facial Mask—Orjene Natural Cosmetics, 1
Clearing Mask—Reviva Labs, 2
Cream Face Mask—Beauty Without Cruelty, Ltd., 1
Daily Scrub/Weekly Mask—Kiss My Face, 1
Dead Sea Black Mud Mask—Reviva Labs, 2
Elder Flower French Clay Masque—Natural
    Bodycare, 1
Face Mask—Dr. Hauschka Cosmetics, 1
Fresh Egg Masque—Borlind of Germany, 2
Fresh Papaya Peel—Zia Cosmetics, 2

Geremy Rose Hot Spring Aloe Mint Mask—New
   Moon Extracts, Inc., 1
Hawaiian Seaweed Face Mask—Reviva Labs, 2
Herb & Clay Mask—Alba Botanica, 1
Icy Mint Peel Off Mask—Natural Solutions, 1
Iris Face Oil—Weleda, Inc., 1
Jojoba Honey Facial Mask—Desert Essence, 1
Jojoba Meal & Oatmeal Facial Mask—Aubrey
   Organics, 1
Kaolin White Clay Mask—Natural Solutions, 1
Light Skin Peel—Reviva Labs, 2
LL Moisturizing Cream Mask—Borlind of
   Germany, 2
LL Vital Cream Mask—Borlind of Germany, 2
Meal 'N Herbs Facial Care Medium—Aubrey
   Organics, 1
Mint Tingle Mask—Earth Science, 1
Mud Mask—Reviva Labs, 2
Mud Masque—Eva Jon Cosmetics, 3
Nature Clay & Mineral Purifying Mask—Avanza
   Corp., 1
Nature's Gate Facial Mask—Levlad, 1
Peaches & Milk Mask—Reviva Labs, 2
Peppermint Arnica Beauty Mask—Paul Penders, 1
Placenta Mask—Reviva Labs, 3
Problem Skin Mask—Reviva Labs, 2
Purifying Clay & Rye Flour Mask—Beauty Without
   Cruelty, Ltd., 1
Rejuvenating Lift Mask—Zia Cosmetics, 2
Sea Mineral Facial Mask—Sea Minerals, 1
Seaclay Seaweed & Clay Beauty Mask—Aubrey
   Organics, 1
Sea-Silk Natural Firming Mask—Natural
   Bodycare, 1
Super Hydrating Mask—Zia Cosmetics, 2

Home Shopping/Mail Order _____

Aloe/Chamomile Masque—Magic of Aloe, 3
Aloe Gel Masque—Jacklyn Cares, 2

Aloe Lift Powder & Aloe Vera Gel—Nutri-Metrics International, Inc., 1

Aloe Moisture Mask—Michel Constantini Natural Cosmetics, 2

Aloe Peel-Off Face Mask—Body Shop, 2

Apache Gold Porcelain Face Mask—Hobe Laboratories, Inc., 1

Aromatherapy Glacial Mud—Ecco Bella, 1

Babyskin Masque—Patricia Allison, 1

Beauty Facial Masque—Magic of Aloe, 3

Chamomile Face Mask—Body Shop, 2

Clarifying Powder Mask—G. T. International, 1

Clay Mask—Michel Constantini Natural Cosmetics, 2

Collagen Moisturizing Mask—Michel Constantini Natural Cosmetics, 3

Cucumber Facial Masque—Key West Fragrance & Cosmetic Factory, Inc., 1

Deep Cleansing Mask with Clay—Kimberly Sayer, Inc., 1

Energizing Face Mud—La Costa Products International, 1

EnVie Facial Masque—Finelle Cosmetics, 2

Firming Mask—Michel Constantini Natural Cosmetics, 2

Gel Mask—Barbizon International, 3

Gel Mask—Shirley Price Aromatherapy, 1

Herbal Facial Masques—Jeanne Rose, 2

Honey Almond Mask—La Costa Products International, 1

Honey & Oatmeal Scrub Mask—Body Shop, 2

Honey Mask—Shirley Price Aromatherapy, 2

Land & Sea Active Mudd—The Face Food Shoppe, 1

Land & Sea Masque/Cleanser—The Face Food Shoppe, 1

Mineral Masque—Nutri-Metrics International, Inc., 1

Miracle Mask—Sombra, 1

Moisture Rich Mask—Mary Kay Cosmetics, Inc., 3

Natural Facial Mask—Jacklyn Cares, 1
Papaya Clarifying Masque—Indian Creek, 1
Peppermint Pickup Stimulation Masque—Patricia
  Allison, 1
Pore Reducer Beauty Treatment Mask—Avon
  Products, Inc., 3
Proto-Refining Masque—Magic of Aloe, 3
Revitalizing Mask—Mary Kay Cosmetics Inc., 3
Sea Scrub Cleaning Mask—Wachter's, 1
Seaweed Masque—The Face Food Shoppe, 1
Sulphur Mask—Michel Constantini Natural
  Cosmetics, 2
Super Mask—Shirley Price Aromatherapy, 1
Swedish Pollenique Purifying Clay Masque—
  Cernitin America, Inc., 1
Translucent Hydrating Facial Mask—La Costa
  Products International, 1
Wheat Germ Masque—The Face Food Shoppe, 1

Specialty Stores

Azu-Gel Soothing Masque—Potions & Lotions, 3
Clay Mask—Body Shop (Calif.), 3
Clay Mask—i Natural Skin Care and Cosmetics, 3
Firming Moisture Masque—Potions & Lotions, 3
Gentle Clay Mask—i Natural Skin Care and
  Cosmetics, 3
Honey Almond Scrub Mask—Body Shop (Calif.), 2
Hydrating Mask—i Natural Skin Care and
  Cosmetics, 3
Mint Scrub Mask—Body Shop (Calif.), 3
Special Clay Masque—Potions & Lotions, 3

FACIAL STEAM AND MIST PRODUCTS

General Retail Stores

Hydrating Mist—Color Me Beautiful, 1

Hydrating Skin Mist—Tyra Skin Care, 2
Natural Rosewater Mist—Cosmyl Cosmetics, 3

Hair Salons

Super Naturals Aloe Vera Skin Spray—North
Country Naturals, 2

Health Food Stores

Aloe-Herb Skin Rejuvenating Mist—Natural
Bodycare, 1
Face Flowers Herbal Facial Steam—Aubrey
Organics, 1
Facial Spray—Reviva Labs, 2
Facial Steam—Alexandra Avery, 1
Facial Steam Bath—Dr. Hauschka Cosmetics, 1
Floral Mist—Aroma Vera, 1
Floral Waters—Aroma Vera, 1
Geremy Rose Rose Tea Mist—New Moon Extracts,
Inc., 1
Herbal Facial Steams—Body Love Natural
Cosmetics, 1
Rose Petals Rosewater—Heritage Store, 1
Rosewater—Home Health Products, 2
Skin Rejuvenating Mist—Earth Science, 1
Sparkling Mineral Water Herbal Mist—Aubrey
Organics, 1

Home Shopping/Mail Order

Elderflower Water—Body Shop, 2
Face & Body Mists—Simplers Botanical Co., 1
Floral Waters—O'Naturel, 1
Floral Waters—Simplers Botanical Co., 1
Honey Water—Body Shop, 2
Hydrating Mist—Finelee Cosmetics, 2
Orange Flower Water—Body Shop, 2
Queen of Hungary's Water—Earth Gifts, 1
Refreshing Moisture Mist—Oriflame
International, 3

Skin Savvy Mist—Strong Skin Savvy, Inc., 1

Specialty Stores ───────────────

Aromatherapy Flowerwater Mist—Bare Escentuals/
Dolphin Acquisition Corp., 2
Facial Sauna Herbs—Scarborough and Company, 1
Herbal Facial Steam—Body Shop (Calif.), 1
Rosewater Mineral Spray—Potions & Lotions, 3
Sea Spray—Body Shop (Calif.), 1

FACIAL TONERS

General Retail Stores ───────────────

Aloe Vera Facial Toner—Botanicus, 1
Ambi Skin Toners— Kiwi Brands Inc., 3
Balancing Tonic—Color Me Beautiful, 1
Botanical Floral Tonic—Smith & Vandiver, Inc., 1
Extra Help Clarifying Toner—TerraNova, 3
Extra-Gentle Refining Toner—Cosmyl Cosmetics, 3
Fragylis Beauty Milk with Flower Blossoms—Jean
Pax, Inc., 1
Fragylis Tonic With Flower Blossoms—Jean Pax,
Inc., 1
Fresh Ups Linen Face Blotters—Andrea
International Ind., 2
Natural Touch Toner—Natural Touch, 3
Purifying Tonic—Color Me Beautiful, 1
Tonic I—Color Me Beautiful, 3

Hair Salons ───────────────

Super Naturals Aloe Vera Skin Toner—North
Country Naturals, 2
Super Naturals Vitamin E Skin Vitalizing Toner—
North Country Naturals, 2
Toner—Aveda Corp., 1

Health Food Stores

Aloe Vera Complexion Toner—Earth Science, 1

Aloegen Rose Hips Tonic Freshener—Levlad, 3

Aloe-Herb Tonic Toner—Natural Bodycare, 1

Arizona Naturals Revitalizing Toner—Arizona Natural Resources, Inc., 2

Blackmores Witch Hazel Toner—Solgar Vitamin Co., 1

Botanee Skin Toner—Reviva Labs, 1

Camphor Lotion—Reviva Labs, 2

Chamomile Angelica Toner—Paul Penders, 1

Derma-aid—Michael's Health Products, 1

Elastin/Collagen Skin Toner—Reviva Labs, 3

Face Lotion—Dr. Hauschka Cosmetics, 1

Flower Essence Toner—Kiss My Face, 1

Gotu Kola Gel—Zia Cosmetics, 1

Grapefruit Refreshing Toner—Orjene Natural Cosmetics, 1

Herbal Facial Toner for Enlarged Pores—Borlind of Germany, 1

Kentucky Clover Toning Mist—Natural Solutions, 3

LL Blossom Dew Gel—Borlind of Germany, 1

N Blossom Lotion—Borlind of Germany, 2

Nature's Gate Herbal Facial Toner—Levlad, 1

Nature's NaPCA & Chamomile Clarifying Skin Toner—Avanza Corp., 1

Orange Blossom Yarrow Toner—Paul Penders, 1

PH Balancer—Eva Jon Cosmetics, 3

Rosa Mosqueta English Lavender Toner—Aubrey Organics

Rose Dew Facial Toner—Borlind of Germany, 1

Sea Kelp Facial Toner—Alba Botanica, 1

Sea Tonic with Aloe Toner—Zia Cosmetics, 1

Serious Skin Daily Toner—Natural Solutions, 1

Skin Toner—Reviva Labs, 2

Skin Tonic—Reviva Labs, 2

Sweet Honey Toning Mist—Natural Solutions, 3

Toner—Alexandra Avery, 1

U Herbal Facial Toner—Borlind of Germany, 1

Violet Rose Skin Toner—Rachel Perry, 3
Z Herbal Facial Toner—Borlind of Germany, 1

Home Shopping/Mail Order

Active Refiner—Shaklee U.S., Inc., 1
Aloe Fresh—Nutri-Metrics International, Inc., 1
Aloe Toner—Michel Constantini Natural
   Cosmetics, 2
Bioflavonoid Toner—Michel Constantini Natural
   Cosmetics, 2
Collagen/Elastin Facial Tonic—Magic of Aloe, 3
Delicate Toner—Oriflame International, 3
Facial Quality Cosmetic Vinegar—Jeanne Rose, 2
Gentle Kelp Toner—Jacklyn Cares, 1
Gentle Refiner—Shaklee U.S., Inc., 1
Hazy-Milk Toner—The Face Food Shoppe., 1
Lavender-Rose Face & Body Toner—The Face Food
   Shoppe, 1
Panthenol Toner—Michel Constantini Natural
   Cosmetics, 2
Protective Toner—Biogime, 1
Rose Toner—The Face Food Shoppe, 1
Rosewater Toner—Michel Constantini Natural
   Cosmetics, 2
Skin Fresh—Nutri-Metrics International, Inc., 1
Skin Toner with Rose & Thyme—Kimberly Sayer,
   Inc., 1
Skin Tonic—Shaklee U.S., Inc., 1
Swedish Pollenique Refreshant Toner—Cernitin
   America, Inc., 1
Toner—World of Aloe, 1
Tonic for Dry Skin—Michel Constantini
   Natural Cosmetics, 2
Toning Lotion—Shirley Price Aromatherapy, 1
Toning Water—La Costa Products International, 1

Specialty Stores

All Herbal Toner—Potions & Lotions, 3

Azulene Toner—Potions & Lotions, 3
Cooling Skin Revitalizer—Scarborough and
    Company, 1
Lemon Mint Toner—Body Shop (Calif.), 3
Mineral Rinse—i Natural Skin Care and
    Cosmetics, 3
Rosewater Toner—Potions & Lotions, 3

FACIAL TREATMENT CREAMS AND LOTIONS

General Retail Stores

Active Performance Hydrating Creme—Cosmyl
    Cosmetics, 3
Ambi Skin Treatments—Kiwi Brands, Inc., 3
Anti-Aging Activator—Clientele, 1
Day/Nite Complex—Tyra Skin Care, 2
Fragylis Revitalizing Cream—Jean Pax, Inc., 3
Hydra-Cell Vital Moisture Creme—Cosmyl
    Cosmetics, 3
Multi-Action Revitalizing Concentrate—Color Me
    Beautiful, 1
*Naturessence Collagen & Elastin Age Control
    Cream*—Avanza Corp., 1
Naturessence European Elastin Firming Facial—
    Avanza Corp., 1
Preventive Age Treatment—Clientele, 1
Surface Refining Lotion—Clientele, 1
Wrinkle Treatment—Clientele, 1

Hair Salons

Atone Skin Rejuvenator—Focus 21, 1
Biokosma Cream Repair—Dr. Grandel, Inc., 1
Biokosma Day Cream—Dr. Grandel, Inc., 1

Day & Night Cream—Dr. Babor Natural
   Cosmetics, 1
Day & Night Protector—Dr. Babor Natural
   Cosmetics, 1
Extra Ordinaire—Jelene International, 3
Hydrovita Soft Lotion—Dr. Babor Natural
   Cosmetics, 1
Lift Powder/Lift Activator—Jelene International, 1
Morning Care—Dr. Babor Natural Cosmetics, 2
pH Balancer—Jelene International, 1
Rejuvene—Mastey de Paris, 1
Revive Creme—Jelene International, 1
Sebum Reducer—Dr. Babor Natural Cosmetics, 1
Silhouette Cream—Dr. Babor Natural Cosmetics, 1

Health Food Stores

Aloegen Elastin Activating Gel—Levlad, 3
Aloe-Herb Skin Rejuvenating Mist—Natural
   Bodycare, 1
Anti U.V. Aging Skin Defense Cream—Avanza
   Corp., 1
Arizona Naturals Original Mud Treatment—
   Arizona Natural Resources, Inc., 2
Azulene Ultra Fleurs Moisture Creme—Natural
   Bodycare, 1
Cellagen Cellular Recovery Creme—Earth Science, 1
Collagen Cream—Reviva Labs, 3
Cream Hydratante—Reviva Labs, 2
Elastin Creme—Orjene Natural Cosmetics, 3
Elastin-Collagen Firming Treatment—Rachel
   Perry, 3
Face Lotion Special—Dr. Hauschka Cosmetics, 1
4 Purpose Cream—Orjene Natural Cosmetics, 3
Gel Stimulante—Reviva Labs, 2
Gel Stimulante w/Trace Minerals—Reviva Labs, 2
Geremy Rose Santa Ana Creme—New Moon
   Extracts, Inc., 1
Ginseng-Collagen Wrinkle Treatment—Rachel
   Perry, 3

Gly-Jene Cellular Creme—Orjene Natural
  Cosmetics, 3
Gold Cream for Wrinkles—Heritage Store, 2
Hi Potency "E" Cellular Treatment—Rachel Perry, 1
LL Regeneration Ampoules—Borlind of Germany, 1
NaPCA Cream—Orjene Natural Cosmetics, 3
Natural Source Vitamin E—Liberty Natural
  Products, 1
Nature Active Nutrisomes Age-Control Cream—
  Avanza Corp., 1
Nature French Formula Firming Facial—Avanza
  Corp., 1
Naturessence Anti U.V. Skin Defense Cream—
  Avanza Corp., 1
Niagara Hydrating Gel with Sodium PCA—Natural
  Solutions, 3
Nourishing Creme—Zia Cosmetics, 3
Oil Regulating Lotion—Beauty Without Cruelty,
  Ltd., 1
Peppermint Melisse Therapeutic Gelee—Paul
  Penders, 1
Placentagen Creme—Earth Science, 1
Preventive Age Cream—Beauty Without Cruelty,
  Ltd., 1
Propolis Derma Cream—Beehive Botanicals, 2
Rose Cream—Dr. Hauschka Cosmetics, 2
Scrub Cream—Beauty Without Cruelty, Ltd., 1
Skin Conditioner N—Dr. Hauschka Cosmetics, 1
Skin Conditioner S—Dr. Hauschka Cosmetics, 1
Skin Cream—Weleda, Inc., 2
Skin Firmer—Reviva Labs, 2
Vit-A Cream—Orjene Natural Cosmetics, 2
Vitamin C Cream—Orjene Natural Cosmetics, 3

Home Shopping/Mail Order

Active-Aloe Total Purpose Cream—Kallima
  International, Inc., 1
Aloe Capture—Magic of Aloe, 3

Orgasea Vitamin E-Avocado Almond Creme—
   Wachter's, 3
Replenishing Formula Wrinkle Cream—Hobe
   Laboratories, Inc., 1
ReVitalizing Isometric Lift Activator—Kallima
   International, Inc., 1
ReVitalizing Isometric Lift Powder—Kallima
   International, Inc., 2
Revitalizing Nutrient Complex—Gruene, 1
Sage & Comfrey Open Pore Cream—Body Shop, 2
Satin Skin—The Face Food Shoppe, 1
Sel-E-RNA Plus "Age-Less" Creme—Nutri-Metrics
   International, Inc., 1
Shine Solution—Avon Products, Inc., 3
Skin Refiner—Avon Products, Inc., 3
Skin Savers Lotion—Key West Fragrance &
   Cosmetic Factory, Inc., 1
Tender Care—Oriflame International, 3
3 Step Non-Surgical Face Lift—Grace Cosmetics/
   Pro-Ma Systems, 2
Ultra Treatment Complex—Oriflame
   International, 3
Un-Line—Patricia Allison, 1
Visible Advantage—Avon Products, Inc., 3

Specialty Stores _____

Botanical Treatment—Potions & Lotions, 3
Collagen Elastin Cream—Body Shop (Calif.), 3
Dermophilic Extract—Potions & Lotions, 3
Jojoba Oil Cream—Body Shop (Calif.), 3
Swiss Collagen Hydrating Complex—Potions &
   Lotions, 3

FEMININE HYGIENE PRODUCTS

Health Food Stores _____

Bee Kind Disposable Douche—Withers Mill
   Company, 2

Douche Concentrate—Bio-Botanica, 1

Home Shopping/Mail Order

Herbal Douche—Jeanne Rose, 2

FINGERNAIL DECORATIONS

General Retail Stores

Nail Jewelry Kit—Ardell International, 2
SuperNail Nail Jewelry & Decorations—Super
Nail, 2

FINGERNAIL REPAIR AND WRAPS

General Retail Stores

Andrea Mendit—Andrea International Ind., 2
Andrea Repair & Wrap Kit—Andrea International
Ind., 2
Easy Mender Nail Kit—Cosmyl Cosmetics, 3
Fiber Nail Wrap—Ardell International, 2
Liquid Silk Nail Wrap—Ardell International, 2
Nail Fix—Delore Nails, 2
Silky Saver Wrap Kit—Ardell International, 2
Super Nails Glue and Wraps—Super Nail, 2
Super Nails Wraps—Super Nail, 2

FINGERNAIL RIDGE FILLERS

General Retail Stores

Andrea Ridgefiller—Andrea International Ind., 2
Ardell Ridge Filler—Ardell International, 2

Silk Ridge Filler—DeLore Nails, 2
Smooth Touch Ridgefiller—Cosmyl Cosmetics, 3

FIRST AID PRODUCTS

Health Food Stores

All-heal Salve—WiseWays Herbals, 2
Aloe Vera First Aid Spray—Jason Natural
   Products, 1
Atomidine—Heritage Store, 1
Bicarbonate of Soda—Heritage Store, 1
Black Walnut-Tea Tree Salve—WiseWays Herbals, 1
Calal Calamine & Aloe for Burns and Bites—Aubrey
   Organics, 1
Casoda for Warts—Heritage Store, 1
Castor Oil—Heritage Store, 1
Castor Oil—Home Health Products, 2
Cimex Lectularius—Heritage Store, 1
Comfrey Salve w/Goldenseal—Autumn Harp, 1
Comfrey-Goldenseal Salve—No Common Scents, 1
De-Tense—Heritage Store, 1
Dragon's Dream (White Balm)—No Common
   Scents, 2
Formula 208—Heritage Store, 3
Formula 545—Heritage Store, 1
Formula 636—Heritage Store, 3
Formula 637—Heritage Store, 3
Glyco-Thymoline—Heritage Store, 1
Iodex—Heritage Store, 1
Itch-aid—Michael's Health Products, 2
Koala Australian Tea Tree Oil—Liberty Natural
   Products, 1
Limewater—Heritage Store, 1
Lithia Water—Heritage Store, 1
Naturetussin—NatureWorks, Inc., 2
NSR Natural Sports Rub—Aubrey Organics, 1

Oak Away—Breezy Balms, 1
PSO Salve—Michael's Health Products, 2
Ray's Liquid—Heritage Store, 1
Ray's Ointment—Heritage Store, 2
Rhus Tox 3—Heritage Store, 1
Scargo—Home Health Products, 2
Similasan for Sore Throats—Similasan Corp., 2
Tea Tree Ointment—Desert Essence, 1
Temple Healer Headache Remedy—Heritage Store, 1
The Original Toothache Aid—Liberty Natural
     Products, 1
Witch Hazel Salve—WiseWays Herbals, 2
WPG Salve—Michael's Health Products, 2

Home Shopping/Mail Order _____

Aloe-Aid Soothing Cream—Grace Cosmetics/
     Pro-Ma Systems, 1
Aloezone—Key West Fragrance & Cosmetic
     Factory, Inc., 1
Alomar—Wachter's, 1
Alpha Remedies—G. T. International, 1
Arniflora Arnica Gel—G. T. International, 1
Bruise Juice—Jeanne Rose, 2
Califlora Calendula Gel—G. T. International, 1
Comfortcaine—Key West Fragrance & Cosmetic
     Factory, Inc., 1
Compound HELP!—Simplers Botanical Co., 1
Ferm-T Natural Antiseptic—Hobe Laboratories,
     Inc., 1
Hercules Oil—Earth Gifts, 1
Injury Oil—Simplers Botanical Co., 1
Key West Medic—Key West Fragrance & Cosmetic
     Factory, Inc., 1
Lipcaine—Key West Fragrance & Cosmetic Factory,
     Inc., 1
Moontime Oil—Simplers Botanical Co., 1
Psoria-Gard—Hobe Laboratories, Inc., 1
Relaxant Creme—Wachter's, 1

Salv-ation Salve—Earth Gifts, 1
Sssssting Stop for Insect Bites—G. T.International, 1

FLEA AND TICK DIPS, REPELLENTS, RINSES, AND SHAMPOOS

General Retail Stores

Flea Off Shampoo—Tender Loving Care Pet Products, 1
Flea Rinse—Tender Loving Care Pet Products, 1

Health Food Stores

Bug-Off—Bug-Off, 1
d-Limonene Citrus Shampoo—Pet Connection, 1
Green Ban Pet Powder—Green Ban, 1
My Pet Aloe Vera Repellent Shampoo—Levlad, 1
Royal Herbal Oil Recharger Concentrate—Pet Connection, 1
Royal Herbal Pet Powder—Pet Connection, 1
Royal Herbal Rechargeable Herbal Collars for Dogs and Cats—Pet Connection, 3

Home Shopping/Mail Order

Cat Herbal Flea Repellent—Jeanne Rose, 2
Dog Herbal Flea Repellent—Jeanne Rose, 2
Herbal Flea Shampoo—Naturally Yours, Alex, 1

FLOOR CLEANERS

General Retail Stores

Earth Rite All-Surface Floor Cleaner—Magic American Corp., 1

Floor and Tile Cleaner—Soap Factory, 1
Holloway House Wood Floor Cleaner & Wax—
    Holloway House, Inc., 1

Health Food Stores _____

Floor Soap—Ecover, 1

FLOOR WAXES

General Retail Stores _____

Holloway House Pure Wax—Holloway House, Inc., 1
Holloway House Quick Shine—Holloway House,
    Inc., 1
Wood Preen—Kiwi Brands, Inc., 3

Home Shopping/Mail Order _____

All Purpose Polish and Wax—AFM Enterprises,
    Inc., 1
Bilo-Floor Wax—Livos Plant Chemistry, 1
One Step Seal and Shine—AFM Enterprises, Inc., 1

FOOT CARE PRODUCTS

General Retail Stores _____

Conditioning Foot Spray—Super Nail, 2
Foot Treat—DeLore Nails, 2
Pedicure Talc—Super Nail, 2
Pumice Stone—Super Nail, 2
Revitalizing Foot Spray—Super Nail, 2

Salon System Cleansing Pedicure Spray—Super Nail, 2
Salon System Pedicure Soak—Super Nail, 2
Stimulating Foot Spray—Super Nail, 2

Hair Salons

Biokosma Golden Feet Bath Salt—Dr. Grandel, Inc., 1
Biokosma Golden Feet Foot Cream—Dr. Grandel, Inc., 1
Biokosma Golden Feet Foot Lotion—Dr. Grandel, Inc., 1

Health Food Stores

Foot Balm—Liberty Natural Products, 2
Foot Cream—Weleda, Inc., 2
FootFriends Foot Deodorant Spray—Liberty Natural Products, 2
Icy Mint Foot Cream—Natural Solutions, 3
Lavilin Foot Deodorant—Micro-Balanced Product Corp., 1
Loofah Anti-Stress Exfoliating Lotion—Jason Natural Products, 1
Pedicure For Athlete's Foot—Heritage Store, 1
Rosemary Foot Balsam—Dr. Hauschka Cosmetics, 1
Rosemary Lotion—Dr. Hauschka Cosmetics, 1
Sage Foot Bath—Dr. Hauschka Cosmetics, 1
Soft & Silky Refreshing Anti-Stress Moisturizing Gel—Jason Natural Products, 1
Sole Soother—Heritage Store, 1
Soothing, Softening, Anti-Stress Foot Soak—Jason Natural Products, 1

Home Shopping/Mail Order

Foot Spray—Golden Pride/Rawleigh, 1
Jojoba Foot Lotion—KSA Jojoba, 1
Peppermint Foot Lotion—Body Shop, 2

FURNITURE POLISHES, WAXES, AND CLEANERS

General Retail Stores

Furniture Magic—Magic American Corp., 1

Holloway House Dust N' Glow—Holloway House, Inc., 1

Holloway House Furniture Cleaner—Holloway House, Inc., 1

Holloway House Lemon Oil—Holloway House, Inc., 1

Holloway House Premium Oil Soap—Holloway House, Inc., 1

Home Shopping/Mail Order

Alis-Furniture Polish—Livos Plant Chemistry, 1

Bekos Bee and Resin Ointment—Livos Plant Chemistry, 2

Dryad-Polish—Livos Plant Chemistry, 1

Furniture Polish—Seventh Generation, 1

Gleivo Liquid Furniture Wax—Livos Plant Chemistry, 2

GERMICIDES

Home Shopping/Mail Order

Basic-G Germicide—Shaklee U.S., Inc., 1

GLASS CLEANERS

General Retail Stores

Clear Vu Glass Cleaner—Clear Vu Products, 1
Glass Mate—Mountain Fresh Products, 1
Nature's Miracle Glass Cleaner—Pets 'N People, 1
Nature's Miracle Plexiglass Cleaner—Pets
   'N People, 1
Sparkle Glass Cleaner—A. J. Funk & Co., 1

Home Shopping/Mail Order

Clean and Green Glass Cleaner—Ecco Bella, 1
Glass Cleaner—Allens Naturally, 1
Glass Cleaner—Seventh Generation, 1

H

HAIR COLORING PRODUCTS

General Retail Stores

Color Comb Instant Color Touch-up—A.I.I.
   Clubman, 1
Egyptian Henna for Black Hair—A.I.I. Clubman, 1
Egyptian Henna for Brown Hair—A.I.I. Clubman, 1
Meta Henna Creme Color—Dena Corp., 1
Meta 1 Step Color—Dena Corp., 1
Streak 'n Tips Temporary Hair Color—A.I.I.
   Clubman, 1

Hair Salons

Color Catalyst—Focus 21, 1
Color-Off—Focus 21, 1
Farmavita Cream Haircolors—Chuckles, Inc., 1
Silhouettes Semi-Permanent Colors—Focus 21, 1
Verocolor—JOICO Laboratories, 1
Verolight—JOICO Laboratories, 1
Veroxide—JOICO Laboratories, 1

Health Food Stores

Color Me Naturally—Paul Penders, 1
Light Touch/Light Mountain Henna—Lotus Light, 1
Rainbow Henna—Rainbow Research Corp., 1

HAIR CONDITIONERS

General Retail Stores

Aloe Honey-Rich Conditioner—Nature de France, 3
Botanical Conditioner—Smith & Vandiver, Inc., 1
Chamomile Protein Conditioner—Nature de
   France, 3
Citre Shine Reconstructing Conditioner—Advanced
   Research Labs, 1
Coconut Creme Conditioner—TerraNova, 3
Curlex Conditioner—Dena Corp., 1
Egyptian Henna Neutral Conditioner—A.I.I.
   Clubman, 1
Egyptian Henna Pre-Mixed Conditioner—A.I.I.
   Clubman, 1
European Mystique Conditioner—Dena Corp., 1
Extra Body Conditioner—SafeBrands, Inc., 1
Extra Gentle Conditioner—SafeBrands, Inc., 1
Faith in Nature Conditioners—Baudelaire, 1
Golden Lotus Herbal Conditioner—Mountain Fresh
   Products, 1

Golden Lotus Lemongrass Conditioner—Mountain
  Fresh Products, 1
Golden Lotus Rosemary & Lavender Conditioner—
  Mountain Fresh Products, 1
Golden Lotus Silk Protein Conditioner—Mountain
  Fresh Products, 1
Lectrify Conditioner—Dena Corp., 1
Meta 1 Step Conditioner—Dena Corp., 1
My Skin Conditioner—Dena Corp., 1
Normal Conditioner—SafeBrands, Inc., 1
Oil-Free Conditioner—Botanicus, 1
Pikaki Conditioner—TerraNova, 3
Rain Conditioner—TerraNova, 3
Smash Conditioner—Dena Corp., 1
Smith & Vandiver Conditioner—Smith & Vandiver,
  Inc., 1
Wysong Rinseless Conditioner—Wysong
  Corporation, 1

Hair Salons

Cherry/Almond Bark Rejuvenating Conditioner—
  Aveda, 1
Excell Silk Fibre Protein Conditioner—Kenra
  Naturals, 1
Farmavita Colorsafe Conditioner—Chuckles, Inc., 1
Paul Mitchell The Conditioner—John Paul Mitchell
  Systems, 1
Super Charged Conditioner—John Paul Mitchell
  Systems, 1
Super Naturals Aloe Vera Conditioner—North
  Country Naturals, 3
Super Naturals Jojoba Conditioner—North Country
  Naturals, 3
Super Naturals Panthenol Conditioner—North
  Country Naturals, 3
Ultra Conditioner for Dry Hair—Zinzare
  International, Ltd., 1

Health Food Stores _____

Alma Conditioner—ShiKai Products, 3
Aloe Vera Conditioner—Jason Natural Products, 2
Aloegen Biogenic Perm Conditioner—Levlad, 3
Aloegen Biogenic Treatment Conditioner—Levlad, 3
Aloegen Biotreatment 22 Conditioner—Levlad, 3
Aloe-Herb Conditioner—Natural Bodycare, 1
*Arizona Naturals Intensive Hair Repair
   Conditioner*—Arizona Natural Resources, Inc., 2
Aurora Henna Conditioner—Aurora Henna, 1
Biotin Conditioner w/NaPCA—Jason Natural
   Products, 2
Blackmores Camomile Conditioner—Solgar Vitamin
   Co., 1
Blackmores Henna & Jojoba Conditioner—Solgar
   Vitamin Co., 1
CamoCare Conditioner—Abkit, Inc., 1
Chamovera Conditioner—Home Health Products, 2
Cherry Bark/Almond Hair Conditioner—Nirvana, 1
Crudoleum Hair Conditioner—Heritage Store, 1
Daily Hair Conditioner—Reviva Labs, 3
E.F.A. Primrose Conditioner—Jason Natural
   Products, 2
Ginkgo Conditioner—Jason Natural Products, 2
GPB Glycogen Protein Balancer Conditioner—
   Aubrey Organics, 2
Hair Conditioner Ampules—Reviva Labs, 3
Hair Conditioning Rinse—Mountain Ocean, 1
Hair Repair Conditioner—Earth Science, 1
Hawaiian Seaweed Hair Conditioner—Reviva
   Labs, 2
Henna & Biotin Conditioner—Rainbow Research
   Corp., 1
Henna Highlights Conditioner—Jason Natural
   Products, 2
Herbal Astringent Conditioner—Earth Science, 1
Herbal Conditioner—Jason Natural Products, 2
Highlighting Conditioner—ShiKai Products, 3

Jojoba & Aloe Rejuvenator & Conditioner—Aubrey Organics, 1

Jojoba Conditioner—Jason Natural Products, 2

Jojoba Lilac Conditioner—Desert Essence, 1

Keratin Conditioner—Jason Natural Products, 2

Lemon Yarrow Cream Rinse Conditioner—Paul Penders, 1

Moisture Replenishing Hair Conditioner—Orjene Natural Cosmetics, 3

Moisturizing Conditioner—Beehive Botanicals, 2

Natural Chamomile Hair Conditioner—Weleda, Inc., 1

Natural Rosemary Hair Conditioner—Weleda, Inc., 1

Nature's Aloe Vera Conditioner—Avanza Corp., 1

Nature's Biotin Plus with Panthenol Conditioner—Avanza Corp., 1

Nature's Gate Herbal Hair Conditioner—Levlad, 1

Nature's Gate Rainwater Herbal Awapuhi Conditioner—Levlad, 1

Nature's Gate Rainwater Herbal Henna Conditioner—Levlad, 1

Nature's Gate Rainwater Rosemary Conditioner—Levlad, 1

Nature's Herbal with Chamomile Conditioner—Avanza Corp., 1

Nature's Jojoba with Panthenol Conditioner—Avanza Corp., 1

Nature's Vege-Protein with Panthenol Conditioner—Avanza Corp., 1

Oil Free Conditioner—Beauty Without Cruelty, Ltd., 1

Oil Free Extra Body Conditioner—Beauty Without Cruelty, Ltd., 1

Oliva Conditioner—Home Health Products, 2

Olive & Aloe Conditioner—Kiss My Face, 1

Oxyfresh Conditioner—Oxyfresh, 1

Polynatural 60/80 Conditioner—Aubrey Organics, 1

Quite Natural Daily Conditioner—Natural
Bodycare, 1
Rosa Mosqueta Conditioning Hair Creme—Aubrey
Organics, 1
Salon Naturals Conditioner—ShiKai Products, 1
Salon Naturals Conditioner Spray On—ShiKai
Products, 3
Sea Kelp Conditioner—Jason Natural Products, 2
Seide Conditioning Rinse—Borlind of Germany, 1
Sonora Desert Jojoba Deep Conditioner—Natural
Solutions, 3
Soy Protein Conditioner—Natural Solutions, 1
Specialized Black Hair Conditioner—Eva Jon
Cosmetics, 3
Stony Brook Botanicals Oil-Free Conditioner—
Rainbow Research Corp., 1
Swimmers Conditioner—Aubrey Organics, 1
Vitamin E Conditioner—Jason Natural Products, 2

Home Shopping/Mail Order

Active-Aloe Hair & Scalp Conditioner—Kallima
International, Inc., 1
Almond Daily Hair Conditioner—Sombra, 1
Almond Intensive Hair Conditioner—Sombra, 1
Aloe Conditioner—Grace Cosmetics/Pro-Ma
Systems, 1
Aromatherapy 60-Second Hair Conditioner—Ecco
Bella, 1
Balancing Conditioner—Mary Kay Cosmetics,
Inc., 3
Banana Conditioner—Body Shop, 2
Botanical Conditioner—Wachter's, 1
Brilliantine Conditioner—Michel Constantini
Natural Cosmetics, 2
Carotene Conditioner—Michel Constantini Natural
Cosmetics, 2

Shaklee Naturals Protective Conditioner—Shaklee U.S., Inc., 1
Shimmer Hair Conditioner—Patricia Allison, 1
Swedish Pollenique Hair Conditioner—Cernitin America, Inc., 1

Specialty Stores

Amino Acid Conditioner with Camomile—Body Shop (Calif.), 3
China Rain Conditioning Rinse—Body Shop (Calif.), 3
Jojoba Oil Conditioner—Body Shop (Calif.), 3
Light Softening Conditioner—Body Shop (Calif.), 3
Nantucket Briar Conditioner—Scarborough and Company, 2
Savannah Garden Conditioner—Scarborough and Company, 2
Spring Rain Conditioner—Scarborough and Company, 2
Sweet Cream Conditioner—Potions & Lotions, 3

HAIR DECORATING PRODUCTS

General Retail Stores

Streaks 'n Tips Glitter Spray—A.I.I. Clubman, 1

HAIR DETANGLERS

Hair Salons

Frehair—Mastey de Paris, 1

Health Food Stores _____

Aloegen Tangle-Free Spray On Conditioner—
Levlad, 3
*Egyptian Henna Hair Rinse—*Aubrey Organics, 1
*Primrose Tangle-Go Lusterizing Spray—*Aubrey
Organics, 2
*Rosemary & Sage Hair & Scalp Rinse—*Aubrey
Organics, 1

Home Shopping/Mail Order _____

*Detangling Spray Conditioner—*Michel Constantini
Natural Cosmetics, 2

HAIR GROWTH PRODUCTS

Health Food Stores _____

*Hair Growth Delay Gel—*Reviva Labs, 2

Home Shopping/Mail Order _____

*Hair Growth Shampoo—*Jeanne Rose, 2

HAIR RELAXERS AND STRAIGHTENERS

General Retail Stores _____

*Curlex Straightener—*Dena Corp., 1
*My Skin Straightener—*Dena Corp., 1
*Smash Relaxer—*Dena Corp., 1

HAIR SPRAY

General Retail Stores

Citre Shine Holding Spritz—Advanced Research Labs, 1

Hair Salons

Changes Hair Spray—Focus 21, 1

Clinch Hair Spray—Paul Mazzotta, Inc., 1

Covert Control Holding Spray—Institute of Trichology, 3

Firm Hold Spray Mist—Zinzare International, Ltd., 1

Fixe Hair Spray—Mastey de Paris, 1

Hair Spray—KMS Research, 1

Liquid Shaper—Sebastian International, Inc., 1

Medium-Hold Spray Mist—Zinzare International, Ltd., 1

Paul Mitchell Fast Drying Sculpting Spray—John Paul Mitchell Systems, 1

Paul Mitchell Freeze and Shine Super Spray—John Paul Mitchell Systems, 1

Paul Mitchell Soft Spray—John Paul Mitchell Systems, 1

Paul Mitchell Super Clean Spray with Awapuhi—John Paul Mitchell Systems, 1

Quick Hold—Paul Mazzotta, Inc., 1

Regide Hair Spray—Mastey de Paris, 1

Splash Hair Spray—Focus 21, 1

Sukesha Maximum Hold Hair Spray—Chuckles, Inc., 1

Sukesha Shape & Style Hair Spray—Chuckles, Inc., 1

Sukesha Styling Hair Spray—Chuckles, Inc., 1

Witch Hazel Hair Spray—Aveda Corp., 1

Health Food Stores

Aloegen Hair Spray—Levlad, 3

Mera Hair Spray—Mera Mattis Beach, 1
Natural Missst Hairspray—Aubrey Organics, 1
Salon Naturals Finishing Spray—ShiKai
 Products, 1
Silk Forte Hair Styling Mist—Earth Science, 1
Silk Lite Hair Styling Mist—Earth Science, 1
Witch Hazel Hair Spray—Nirvana, 1

Home Shopping/Mail Order

Aloe Finish Hair Spray—Key West Fragrance &
 Cosmetic Factory, Inc., 1
Arrange Mist Hair Spray—Shaklee U.S., Inc., 1
Hair Lover's Ferm-Hold Hair Spray—Hobe
 Laboratories, Inc., 1
Hair Spray—Jacklyn Cares, 2
Hold Absolute Hair Spray—La Costa Products
 International, 1
Holding Mist—Finelle Cosmetics, 2
Styling and Finishing Hair Spray—Michel
 Constantini Natural Cosmetics, 2

HAIR TREATMENTS

General Retail Stores

Egyptian Henna Hot Oil Treatment—A.I.I.
 Clubman, 1
Gee, Your Hair Smells Terrific—Andrew Jergens, 3
Jojoba Super Conditioning Treatment—A.I.I.
 Clubman, 1
Nourishing Rinse & Revitalizer—Clientele, 1
Purifying Shine Rinse—Clientele, 1

Hair Salons

Basic Superpac—Mastey de Paris, 1

Paul Mitchell Awapuhi Moisture Mist—John Paul Mitchell Systems, 1

Paul Mitchell Hair Repair Treatment—John Paul Mitchell Systems, 2

Paul Mitchell Seal and Shine—John Paul Mitchell Systems, 2

Phine'—JOICO Laboratories, 1

pHinish—KMS Research, 1

Pre Wrap—L'anza, 1

Pree—L'anza, 1

Prolimin Gold—KMS Research

Protege Hair Sunscreen—Mastey de Paris, 1

Re Balance—L'anza, 1

Re-Building Formula—Focus 21, 1

Re-Moisturizing Treatment—Focus 21, 1

Reconditioning Formula—Focus 21, 1

RePair—KMS Research, 1

RePlace—KMS Research, 1

Rosemary/Mint Equalizer—Aveda Corp., 1

SeaPlasma Hair Booster—Focus 21, 1

SeaPlasma Spray Moisturizer—Focus 21, 1

Sheen—Sebastian International, Inc., 1

Silker—KMS Research, 1

Sukesha Conditioning Rinse—Chuckles, Inc., 1

Sukesha Hair Moisturizing Treatment—Chuckles, Inc., 1

Sukesha Shine and Body—Chuckles, Inc., 1

Super Naturals Aloe Vera Intensive Hair Conditioner—North Country Naturals, 3

Trilogy—KMS Research, 1

Ultra-Pak—KMS Research, 1

Health Food Stores

Aloegen Biotreatment 22 Revitalizing Gel—Levlad, 3

Arizona Naturals Daily Moisturizing Creme Rinse—Arizona Natural Resources, 2

Crowning Glory Hair Creme—WiseWays Herbals, 1

Crudoleum Hair Rinse—Heritage Store, 1

Crudoleum Hair Treatment—Heritage Store, 1

E.F.A. Primrose Protein Pak—Jason Natural
   Products, 2
German Herbal Hair Repair—Paul Penders, 1
Ginko Leaf/Ginseng Root Moisturizing Jelly—
   Aubrey Organics, 1
Ginseng Hair Control—Aubrey Organics, 1
Golden Apple Cider Vinegar Hair Rinse—WiseWays
   Herbals, 1
HA-5 Deep Conditioning Pack—Earth Science, 1
Hair Conditioning Treatment—Mountain Ocean, 1
Hair Lotion with Rosemary—Weleda, Inc., 1
Hair Mask for Moisture Balance—Reviva Labs, 3
Herbal-Satin Natural Hair Revitalizer—Natural
   Bodycare, 1
Mera CPR—Mera Mattis Beach, 1
Quite Natural Intensive Reconstuctive Pac—Natural
   Bodycare, 1
Rainbow Vegetable Reconstructor—Rainbow
   Research Corp., 1
Raven Apple Cider Vinegar Hair Rinse—WiseWays
   Herbals, 1
Rosemary Hair Oil—Weleda, Inc., 1
Rosemary/Mint Hair Rinse—Nirvana, 1
Seide Revitalizer—Borlind of Germany, 1

Home Shopping/Mail Order

Cream Rinse with Jojoba—KSA Jojoba, 1
Energizer Hair Follicle Stimulator—Hobe
   Laboratories, Inc., 1
Hair Moisturizing Treatment—Finelle Cosmetics, 2
Henna Treatment Wax—Body Shop, 2
Herbal Hot Oil Treatment—Jeanne Rose, 2
Lavender Intensive Conditioning Pack—
   O'Naturel, 1
Natural Lustre Finishing Rinse—Nutri-Metrics
   International, 1
Protein Creme Rinse—Body Shop, 2

Restorative Hair Mud—La Costa Products
  International, 1
Rosemary Hair Oil—Jeanne Rose, 2
Rosemary-Jojoba Hair Oil—Jeanne Rose, 2
Shiny Hair Clean Scalp Tonic—Ecco Bella, 1
Soft 'N Silky—Granny's Old Fashioned Products, 1
Softique Hair Re-Moisturizer—Magic of Aloe, 3

Specialty Stores

Leave On Conditioning Spray—Body Shop (Calif.), 3
Lemon Cream Rinse—Body Shop (Calif.), 3
Protein Conditioning Pack—Body Shop (Calif.), 3
Protein Cream Rinse—Potions & Lotions, 3

HAIRBRUSHES

Health Food Stores

Fuchs Hairbrushes—Auro Trading Company, 3

HAIRSTYLING AIDS

General Retail Stores

Citré Shine Liquid Relector—Advanced Research
  Labs, 1
Citré Shine Shaping Gel—Advanced Research
  Labs, 1
Citré Shine Shine Miracle—Advanced Research
  Labs, 1
Curl 'n Set Gel—A.I.I. Clubman, 1
Curl 'n Set—A.I.I. Clubman, 1

Super Set Lotion—A.I.I. Clubman, 1
World's Greatest Hair Styling Gels—Avanza Corp., 1

Hair Salons

Aerogel—Institute of Trichology, 3
Anti-Humectrant Shine—Aveda Corp., 1
Bream—Paul Mazzotta, Inc., 1
C'est La Force—KMS Research, 1
Conditioning Fixx—KMS Research, 1
Construx Building Gel—Paul Mazzotta, Inc., 1
Construx Building Spray—Paul Mazzotta, Inc., 1
Control—KMS Research, 1
Control and Finishing Mist—Institute of
    Trichology, 3
Creme de la Mousse—Regular—KMS Research, 1
Curl Activator—KMS Research, 1
Defining Creme—KMS Research, 1
Designer—Mastey de Paris, 1
Dramatic FX—L'anza, 1
Empower Spray Gel—Kenra Naturals, 1
Enplace—Mastey de Paris, 1
Enviro-Tek Spray Gel—Focus 21, 1
Fashion Styling Mousse—Institute of Trichology, 3
Fixcream—Paul Mazzotta, Inc., 1
Fizz Extra—Sebastian International, Inc., 1
Flax Seed/Aloe Sculpting Gel—Aveda Corp., 1
Flax Seed/Aloe Spray-on Gel—Aveda Corp., 1
Fomme Sprae—L'anza, 1
Gelato—KMS Research, 1
Glazer Gel—L'anza, 1
Hair Candy Mousse—Focus 21, 1
Hair Toys Thikk—Focus 21, 1
Hair Toys Wave Gel—Focus 21, 1
Hairforce Styling/Finishing Spray—Kenra
    Naturals, 1
HairHold—KMS Research, 1
Humectant Shine—Aveda Corp., 1
I.C.E.—JOICO Laboratories, 1
I.C.E. Gel—JOICO Laboratories, 1

Sukesha Freeze Frame Super Spray—Chuckles, Inc., 1

Sukesha Sculpturing Lotion—Chuckles, Inc., 1

Sukesha Styling Gel—Chuckles, Inc., 1

Sukesha Styling Mousse—Chuckles, Inc., 1

Super Naturals Panthenol Styling Gel—North Country Naturals, 2

Super Quatre—L'anza, 1

Transformations Light Pomade—JOICO Laboratories, 1

Transformations Liquid Polish—JOICO Laboratories, 1

Transformations Spray Gel—JOICO Laboratories, 1

Trav'allo—JOICO Laboratories, 1

Health Food Stores

Aloegen Hair Sculpting Gel—Levlad, 3

Aloegen Styling Spritz—Levlad, 3

Arizona Naturals Hair Gel—Arizona Natural Resources, Inc., 2

Arizona Naturals Quick Dry Finishing Spray—Arizona Natural Resources, Inc., 2

B5 Design Gel—Aubrey Organics, 1

Chestnut Brown Body Highliter Mousse—Aubrey Organics, 1

Flax/Aloe Vera Spray Gel—Nirvana, 1

Flax Sculpturing Gel—Nirvana, 1

Golden Chamomile Highliter Mousse—Aubrey Organics, 1

Hair Cream—Orjene Natural Cosmetics, 3

Mera Misting Gel—Mera Mattis Beach, 1

Mera Sculpting Gel—Mera Mattis Beach, 1

More-Curl—Natural Bodycare, 1

Non-Aerosol Mousse—Earth Science, 1

Perfect Hair Dress—Orjene Natural Cosmetics, 2

Protein Hair Thickener—Natural Solutions, 3

Quite Natural Styling Gel—Natural Bodycare, 1

Salon Naturals Styling Gel—ShiKai Products, 1

Salon Naturals Styling Spray—ShiKai Products, 1

Sculpting Glaze—Earth Science, 1
Seide Styling Lotion—Borlind of Germany, 1
Soft Black Body Highliter Mousse—Aubrey
    Organics, 1
Vitamin B Enriched Hair Spritz—Natural
    Solutions, 1
Vitamin B Enriched Styling Gel—Natural
    Solutions, 3

Home Shopping/Mail Order

Aloe Hair Gel—Body Shop, 2
Crystal Clear Sculpting Spray Gel—La Costa
    Products International, 1
Enriched Styling Gel—Mary Kay Cosmetics, Inc., 3
Finishing Spray—Mary Kay Cosmetics, Inc., 3
Hair Gel—Body Shop, 2
Hair Lover's Hair Teaser Styling Spray—Hobe
    Laboratories, Inc., 1
Hair Moisturizing Treatment—Finelle Cosmetics, 2
Hair Sculpting Gel with Citrus—Ecco Bella, 1
Hair Texturizer—Wacheter's, 1
Liquid Mousse—Finelle Cosmetics, 2
Liquid Mousse—Finelle Cosmetics, 2
Misting Gel—Zinzare International, Ltd., 1
Pure Shine—Michel Constantini Natural
    Cosmetics, 2
Reflectives Brilliant Shine Spray—La Costa
    Products International, 1
Sculpting and Styling Gel—Michel Constantini
    Natural Cosmetics, 2
Slick Hair Styler—Body Shop, 2
Styling Gel—Zinzare International, Ltd., 1
Ultimate Styling Mousse—Mary Kay Cosmetics,
    Inc., 3
Uplifting Spray Gel—Michel Constantini Natural
    Cosmetics, 2
Vitagel Sculpting Gel—La Costa Products
    International, 1

Specialty Stores ────────────────────

Styling Gel—Body Shop (Calif.), 3

HAND AND BODY LOTIONS

General Retail Stores ────────────────

Age Defense Hand and Body Lotion—Clientele, 1
Aloe & E All Over Lotion—Colonial Dames Co.,
Ltd., 1
Aloe & E & PABA Complex—Colonial Dames Co.,
Ltd., 1
Aloe & Lanolin Skin Conditioning Lotion—Andrew
Jergens, 3
Aloe Lite Lotion—SafeBrands, Inc., 1
Aloe Vera All Over Lotion—Colonial Dames Co.,
Ltd., 1
Apiana Honey Body Lotion—Baudelaire, 2
Apricot Body Lotion—TerraNova, 3
Chamomile Dry Skin Lotion—Nature de France, 3
China Lily Body Lotion—TerraNova, 3
China Mist Body Lotion—TerraNova, 3
De Maris Body Lotion—Cosmyl Cosmetics, 3
Eversoft Concentrated Skin Care Lotion—Andrew
Jergens, 3
Exclusive Hand Lotion—Tyra Skin Care, 2
Fine Toiletries Hand & Body Lotion—Smith &
Vandiver, Inc., 1
Fragylis Hand Beauty Cream—Jean Pax, Inc., 1
Golden Lotus Ginseng-Aloe Moisturizing Lotion—
Mountain Fresh Products, 1
Golden Lotus Jojoba-E Moisturizing Lotion—
Mountain Fresh Products, 1
Good Clean Fun Children's Lotion—Smith &
Vandiver, Inc., 1

Hand Cream—Colonial Dames Co., Ltd., 1

Hand/Body Lotion No Fragrance—Tyra Skin Care, 2

Jacob Hooy Skin Cream—Baudelaire, 2

Jasmine Body Lotion—TerraNova, 3

Jergens All Purpose Face Cream—Andrew Jergens, 3

Jergens Dry Skin Lotion—Andrew Jergens, 3

Jergens Extra Dry Skin Lotion—Andrew Jergens, 3

Long Lasting Extra Dry Skin Lotion—SafeBrands, Inc., 1

Moon Shadows Body Lotion—TerraNova, 3

Multi-Vitamin Dry Skin Lotion—SafeBrands, Inc., 1

Naturessence Collagen & Herbal Protective Hand Creme—Avanza Corp., 1

Naturessence Swiss Collagen Body Creme—Avanza Corp., 1

Oil of Jojoba Lotion—Colonial Dames Co., Ltd., 1

Oil-Free Hand & Body Lotion—Botanicus, 1

Paradise Mist Body Lotion—TerraNova, 3

Pikaki Body Lotion—TerraNova, 3

Rain Body Lotion—TerraNova, 3

Silk Skin Hand Creme—Cosmyl Cosmetics, 3

Silk Skin Hand Lotion—Cosmyl Cosmetics, 3

Sinclair & Valentine Hand Body Lotion—Smith & Vandiver, Inc., 1

Tropical Gardenia Body Lotion—TerraNova, 3

Vitamin E & Lanolin Skin Conditioning Lotion—Andrew Jergens, 3

Vitamin E Body Lotion—Colonial Dames Co., Ltd., 1

Vitamin E Cream—Colonial Dames Co., Ltd., 1

Vitamin E Lotion With Cream—Colonial Dames Co., Ltd., 1

Vitamin E Oil—Colonial Dames Co., Ltd., 1

Hair Salons

Biokosma Body Lotion—Dr. Grandel, Inc., 1

Biokosma Cleopatra Body Lotion—Dr. Grandel, Inc., 2

Biokosma Lemon Hand Cream—Dr. Grandel, Inc., 1

Boikosma Limetta Hand Cream—Dr. Grandel, Inc., 1
Biokosma Sandalwood Body Lotion—Dr. Grandel, Inc., 1
Bodytone Lotion—Jelene International, 1
Super Naturals Aloe Vera Body Lotion—North Country Naturals, 2
Super Naturals Elastin Lotion—North Country Naturals, 3
Super Naturals Vitamin E Body Lotion—North Country Naturals, 2
Topical Nutrition Hand and Body Lotion—Institute of Trichology, 3
Velvet-Peach Hand Cream—House of Lowell, Inc., 3
Velvet-Peach Hand Lotion—House of Lowell, Inc., 3
Wind 'n Weather Hand Creme—House of Lowell, Inc., 3

Health Food Stores

All Day Olive & Aloe Protective Moisture Creme—Kiss My Face, 1
All Purpose Skin Lotion with E—Amsource International, 2
Almond Glow—Home Health Products, 2
Aloe "E" Lotion—Rachel Perry, 1
Aloe/Rose Body Lotion—Autumn Harp, 1
Aloe Vera Gel Hand & Body Lotion—Jason Natural Products, 1
Apricot Hand & Body Lotion—Jason Natural Products, 1
Arizona Naturals All Over Body Lotion—Arizona Natural Resources, Inc., 2
Aura Glow—Heritage Store, 2
Avocado Hand Creme—Orjene Natural Cosmetics, 3
Blackmores Calendula Hand Creme—Solgar Vitamin Co., 3
Blackmores Evening Primrose Body Lotion—Solgar Vitamin Co., 1
Body Balm—Borlind of Germany, 2
Body Milk—Dr. Hauschka Cosmetics, 1

Botanee Hand & Body Lotion—Reviva Labs, 1
Calming Flower Body Lotion—Paul Penders, 1
CamoCare Hand & Body Lotion—Abkit, Inc., 1
Cocoa Butter Hand & Body Lotion—Jason Natural
　　Products, 1
Cocoa Queen Hand Lotion—WiseWays Herbals, 2
Collagen & Almond Hand & Body Creme—Aubrey
　　Organics, 3
E.F.A. Primrose Hand & Body Lotion—Jason
　　Natural Products, 1
84% Aloe Body Moisturizing Lotion—Jason Natural
　　Products, 1
Elastin Body Lotion—Reviva Labs, 3
Eva Jon All Purpose Body Cream—Eva Jon
　　Cosmetics, 3
Eva Jon Hand Creme—Eva Jon Cosmetics, 3
Evening Primrose Complexion & Body Lotion—
　　Aubrey Organics, 1
Flowers of Lilac Hand Lotion—Beauty Without
　　Cruelty, Ltd., 1
Geremy Rose Sweet Almond Body Creme—New
　　Moon Extracts, Inc., 1
Golden Moisturizing Oil—Rainbow Research
　　Corp., 2
Hand Balm—Borlind of Germany, 2
Hawaiian Seawood Body Lotion—Reviva Labs, 2
Herbal Creme—Jason Natural Products, 1
Herbal Hand & Body Lotion—Jason Natural
　　Products, 1
Herbal Mantle Hand and Body Lotion—Natural
　　Bodycare, 1
Honey & Calendula Moisture Creme—Kiss My
　　Face, 1
Hydrating Hand & Body Lotion—Earth Science, 1
InterCell Hand & Body Lotion—Reviva Labs, 3
Jojoba Rosemary Lotion—Desert Essence, 1
Keoki Papaya Hand and Body Cream—Natural
　　Solutions, 3

Stony Brook Botanicals Oil-Free Body Lotion—
  Rainbow Research Corp., 1
Truly Moist Extra Dry Skin NaPCA Lotion—Desert
  Naturels, 1
Truly Moist Original Formula NaPCA Lotion—
  Desert Naturels, 1
Very Emollient Body Lotion—Alba Botanica, 1
Vitamin A & E Moisture Creme—Kiss My Face, 1
Vitamin E Hand & Body Lotion—Jason Natural
  Products, 1
Vitamin E Skin Lotion—Home Health Products, 2

Home Shopping/Mail Order

Active-Aloe Hand & Body Lotion—Kallima
  International, Inc., 1
Aloe Body Lotion—Gruene, 1
Aloe Hand and Body Lotion—Grace Cosmetics/
  Pro-Ma Systems, 1
Aloe Lotion—Body Shop, 2
Aloe Vera Moisture Cream—Body Shop, 2
Apricot Hand & Body Lotion—Shaklee U.S., Inc., 1
Avocado Body Lotion—Michel Constantini Natural
  Cosmetics, 2
Body Silk—Oriflame International, 3
Body Soft—Oriflame International, 3
Cocoa Butter Hand & Body Lotion—Body Shop, 2
Collagen Hand Cream—Oriflame International, 3
Deep Skin Hand and Body Lotion—Hobe
  Laboratories, Inc., 1
Dewberry 5 Oils Lotion—Body Shop, 2
Exquisite Body Lotion—Mary Kay Cosmetics, Inc., 3
Glycerin & Rosewater Lotion with Vitamin E—Body
  Shop, 2
Hand & Body Lotion—Magic of Aloe, 3
Hand and Body Lotion—Jacklyn Cares, 2
Hand and Body Lotion—Key West Fragrance &
  Cosmetic Factory, Inc., 1
Hand and Body Lotions—Michel Constantini
  Natural Cosmetics, 2

Hand Lotion—Shirley Price Aromatherapy, 1
Hawthorn Hand Cream—Body Shop, 2
Herbal Blend Multi-Purpose Cream—Shaklee U.S., Inc., 1
Herbal Moist Body Lotion—Earth Gifts, 1
Irish Legend Hand Beauty—Patricia Allison, 1
Jojoba Body Cream—KSA Jojoba, 1
L'Arome Echoes Body Lotion—L'Arome USA, Inc., 1
Lady Shaklee Body Creme—Shaklee U.S., Inc., 1
Moisture Guard—Granny's Old Fashioned Products, 1
Moisture Rich Hand & Body Lotion—Nutri-Metrics International, Inc., 1
Moisturizing Hand Creme—Wachter's, 1
Petalskin Hand and Body Balm—Patricia Allison, 1
Pineapple Coconut Hand & Body Lotion—Sombra, 1
Sel-E-RNA Plus Anti-Aging Body Lotion—Nutri-Metrics International, Inc., 1
Shaklee's Naturals All-Over Moisturizer—Shaklee U.S., Inc., 1
Skin Moist—Key West Fragrance & Cosmetic Factory, Inc., 1
Skin Savvy Cream—Strong Skin Savvy, Inc., 2
Softa Skin Hand & Body Lotion—Wachter's, 1
Soothing Lotion—Shirley Price Aromatherapy, 1
Vitamin E Hand & Body Lotion—Bronson Pharmaceuticals, 3
Vitamin E-D-A Body Lotion—La Costa Products International, 2
Youthening Hand & Body Moisture—Irma Shorell, 2

Specialty Stores

A.D.E. Lotion—Potions & Lotions, 3
ADE Light Lotion—Bare Escentuals/Dolphin Acquisition Corp., 2
Aloe & Comfrey Lotion—Bare Escentuals/Dolphin Acquisition Corp., 2
Aloe Vera Lotion—Potions & Lotions, 3

Apricot Cocoa Butter Lotion—Potions & Lotions, 3
Avocado Lotion—Crabtree & Evelyn, Ltd., 1
Bee Pollen Lotion—i Natural Skin Care and
    Cosmetics, 3
Bee Pollen Moist—i Natural Skin Care and
    Cosmetics, 3
Buttermilk Moisturizing Cream—Crabtree &
    Evelyn, Ltd., 2
China Rain Lotion—Body Shop (Calif.), 3
Collagen-Elastin Lotion—Potions & Lotions, 3
Cucumber Lotion—Crabtree & Evelyn, Ltd., 1
French Vanilla Lotion—Potions & Lotions, 3
French Vanilla X-Rich Lotion—Bare Escentuals/
    Dolphin Acquisition Corp., 2
Glycerin & Rose Lotion with Vitamin E—Body Shop
    (Calif.), 3
Glycerin & Rosewater—Body Shop (Calif.), 3
Goatmilk Moisturizing Cream—Crabtree & Evelyn,
    Ltd., 2
Hand & Body Lotion—Potions & Lotions, 3
Hand Cream With Vitamin E—Body Shop (Calif.), 3
Jojoba Day Cream—Crabtree & Evelyn, Ltd., 1
Meadowfoam Lotion—Body Shop (Calif.), 1
Moisture Lotion—Body Shop (Calif.), 3
Perfumed Hand & Body Lotion—Scarborough and
    Company, 2
Rosewater & Glycerin Lotion—Bare Escentuals/
    Dolphin Acquisition Corp., 2
Rosewater Lotion—Crabtree & Evelyn, Ltd., 1
Scented Vitamin E Lotion—Bare Escentuals/
    Dolphin Acquisition Corp., 2
Spectrum Lotion—Potions & Lotions, 3
Sunscreen Lotion—Potions & Lotions, 3
Unscented Vitamin Cream—Body Shop (Calif.), 3
Wheatgerm Moisturizing Cream—Crabtree &
    Evelyn, Ltd., 1
Wild Rose & Meadowsweet Lotion—Crabtree &
    Evelyn, Ltd., 1

HEMORRHOID TREATMENTS

General Retail Stores ───────────────

Lanex—Carma Laboratories, Inc., 2

Health Food Stores ───────────────

Homeopathic Hemorrhoid Cream—Bio-Botanica, 1
Key-E Suppositories—Carlson Laboratories, 1
Rectal-Aid—Michael's Health Products, 2
TIM for Hemorrhoids—Heritage Store, 3

HERBAL EXTRACTS AND REMEDIES

General Retail Stores ───────────────

Botanicals Mood Spray—Smith & Vandiver, Inc., 1

Hair Salons ───────────────

Herbal Extract Recovery Gel—Jurlique D'Namis Ltd., 1

Health Food Stores ───────────────

Alive Energy Assorted Herbal Formulations—Auro Trading, 1
Alive Energy Body Gain—Auro Trading, 3
Alive Energy Full Spectrum Energy—Auro Trading, 3
Amrita Chyavanprash—Auromere, Inc., 1
Analgesic Yarrow Liniment—Lakon Herbals, 2
Botanicaps—Bio-Botanica, Inc., 3
Comfrey Herbal Salve—Lakon Herbals, 2

Dr. Pati's Ayurvedic—Auromere, Inc., 1
*Dragon Eggs Chinese Herbal Energy
    Formulations*—Auro Trading, 1
Essential Care Skin Cream—Lakon Herbals, 1
Fluid Extracts—WiseWays Herbals, 1
Goldenseal Comfrey Salve—Lakon Herbals, 2
Herbal Balm—Alexandra Avery, 1
Herbal Savvy Salves—Country Comforts, 2
Homeopathic Products—Bio-Botanica, Inc., 1
Homeopathic Topical Creams—Bio-Botanica, Inc., 1
Marigold Cream—NatureWorks, 1
Nature's Alchemy Herbalmune Herbal Extract—
    Lotus Light, 1
Nature's Answer Extracts—Bio-Botanica, Inc., 1
Oil of Hypericum—Lakon Herbals, 1
Passion Flower Fusion Nervine—Heritage Store, 1
Peelu Extract—Peelu Products, 1
Pure Apple Cider Vinegar Extracts—WiseWays
    Herbals, 1
Swedish Bitters—NatureWorks Inc., 1
Swedish Herbs—NatureWorks, Inc., 1
Transformational Healing Blends—WiseWays
    Herbals, 1

Home Shopping/Mail Order

Blu-Green Manna—Hobe Laboratories, Inc., 1
Col-S-Rol Tea—Hobe Laboratories, Inc., 1
Face Herbs—The Face Food Shoppe, 1
Formulas for Health—Golden Pride/Rawleigh, 1
Hair Herbs—The Face Food Shoppe, 1
Health and Home Remedies—Golden Pride/
    Rawleigh, 1
Herbal Salve—Heavenly Soap, 2
Herbs for Health—Golden Pride/Rawleigh, 1
Liquid Herbal Compounds—Simplers Botanical
    Co., 1
PMS Time—Hobe Laboratories, Inc., 1

Simplers Liquid Herbal Concentrates—Simplers
   Botanicals, 1
Treasure of the Incas—Hobe Laboratories, Inc., 1
Vita Green—Hobe Laboratories, Inc., 1

INCENSE

General Retail Stores

Incense Reeds—TerraNova, 1

Health Food Stores

Auromere Ayurvedic Incense of India—Auromere,
   Inc., 1
Auroshikha Incense—Auroma International, 1
Centenary Incense—Auroma International, 1
Herbal Vedic Incense—Auroma International, 1
Incense—No Common Scents, 1
Sri Aurobindo Ashram Incense—Auromere, Inc., 1

Specialty Stores

Incense—Potions & Lotions, 3
Incense—Scarborough and Company, 1

INSECT REPELLENTS

Health Food Stores _____

Bug-Off—Bug-Off, 1
Bygone Bugs—Lakon Herbals, 1
Green Ban for People—Green Ban, 1
Insect Repellent—No Common Scents, 1
Insect Repellent Oil—Aroma Vera, 1
Shoo—Michael's Health Products, 1
SWAT Insect Deterent—Liberty Natural Products, 1
SWAT Lotion—Liberty Natural Products, 1

Home Shopping/Mail Order _____

Cembra Essence—Livos Plant Chemistry, 1
Potpourri Moth & Bug Repellent—Jeanne Rose, 2

J

JOJOBA OIL

General Retail Stores _____

Golden Lotus Pure Jojoba Oil—Mountain Fresh
Products, 1

Hair Salons _____

Super Naturals Pure Jojoba Oil—North Country
Naturals, 1

Health Food Stores

Jojoba Oil—Aroma Vera, 1
Jojoba Oil—Desert Essence, 1
Jojoba Oil 100%—Aubrey Organics, 1
Jojoba Oil—100% Pure—Jason Natural Products, 1
Natural Jojoba Oil—Earth Science, 1

Home Shopping/Mail Order

Jojoba Oil—Bronson Pharmaceuticals, 1
Jojoba Oil—KSA Jojoba, 1
Jojoba Oil—Body Shop, 2
100% Pure Jojoba Oil—Hobe Laboratories, Inc., 1

K

KITCHEN AND COUNTER CLEANERS

General Retail Stores

Countertop Magic—Magic American Corp., 1
Earth Rite Countertop Cleaner-Polish—Magic American Corp., 1
Kitchen Magic—Magic American Corp., 1
Shiny Sinks Plus Cleaner—SerVaas Laboratories, 1

L

LAUNDRY PRE-WASHES AND SPOT REMOVERS

General Retail Stores

Laundry Stain Remover—Soap Factory, 1
Rit Grease, Soil and Stain Remover—Best Foods, 1
Winter White Pre-Wash—Mountain Fresh
Products, 1

Health Food Stores

Golden Lotus Pre-Spot—Lotus Light, 1

Home Shopping/Mail Order

Soil Away—Granny's Old Fashioned Products, 1

LAUNDRY SOAPS AND DETERGENTS

General Retail Stores

Baby Laundry Detergent—Soap Factory, 1
Liquid Laundry Detergent—Soap Factory, 1
Winter White Liquid Laundry Detergent—Mountain
Fresh Products, 1

Winter White Powdered Detergent—Mountain Fresh
   Products, 1

Health Food Stores

Golden Lotus Laundry Detergent—Lotus Light, 1
Laundry Powder—Ecover, 1
Liquid Laundry—Ecover, 2
New America Clean Laundry Detergent—Abkit,
   Inc., 1
Oxyfresh Laundry Detergent—Oxyfresh, 1
Premium Laundry Liquid—Sierra Dawn, 1

Home Shopping/Mail Order

Basic-L Laundry Concentrate—Shaklee U.S., Inc., 1
Clean and Green—Ecco Bella, 1
Double-Strength Laundry Liquid—Seventh
   Generation, 1
Improved Liquid-L Laundry Concentrate—Shaklee
   U.S., Inc., 1
Laundro-Kleen—Wachter's, 1
Liquid Laundry Detergent—Allens Naturally, 1
Liquid Laundry Detergent with Lemon—Ecco
   Bella, 1
Neway Laundry & Cleaning Compound—Neway, 1
Newliquid Laundry Detergent—Neway, 1
Power Plus—Granny's Old Fashioned Products, 1
Professional Laundry Compound Powder—Home
   Service Products, 1
Professional Liquid Laundry Detergent—Home
   Service Products, 1

LAXATIVES

Health Food Stores

Innerclean Laxative—Heritage Store, 1

Sulflax—Heritage Store, 1
Zilatone—Heritage Store, 3

Home Shopping/Mail Order

Colon Guard—Hobe Laboratories, Inc., 1

LIGHTERS, DISPOSABLE

General Retail Stores

Cricket Disposble Lighters—Wilkinson Sword, 1
Feudor Disposable Lighters—Wilkinson Sword, 1

LIME DEPOSIT REMOVERS

General Retail Stores

Power House Lime Remover—SerVaas
Laboratories, 1

LIP BALMS

General Retail Stores

Carmex—Carma Laboratories, Inc., 2
Good Clean Fun Lip Balm—Smith & Vandiver,
Inc., 1
Lip Renewal Creme—Andrea International Ind., 2
LipSaver—Andrea International Ind., 2

Pink Blush PABA Lip Gloss—TerraNova, 3
Rose Blush PABA Lip Gloss—TerraNova, 3
Sheer Natural PABA Lip Balm—TerraNova, 3

Health Food Stores

Apricot Aloe Lip Balm—Alexandra Avery, 1
Aura Glow Lip Balm—Heritage Store, 2
E-Gem Lip Care—Carlson Laboratories, 3
For Lips Moisturizer—Borlind of Germany, 2
Heavenly Scents Lip Balms—Liberty Natural
    Products, 1
Jojoba Aloe Vera Lip Balm—Desert Essence, 1
Lip Balm—Borlind of Germany, 2
Lip Balm—Weleda, Inc., 2
Lip Cream—Dr. Hauschka Cosmetics, 2
Lip Creme Sticks—Country Comfort, 2
Lip Lover—Rachel Perry, 2
Lip Protector and Moisturizer—Earth Science, 1
Lip Sense—Autumn Harp, 1
Lip Trip—Mountain Ocean, 2
Natural Lip Balms—WiseWays Herbals, 2
Palma Christi—Home Health Products, 2
Propolis Lip Balm—Beehive Botanicals, 2
Tea Tree Lip Balm—Desert Essence, 1
Ultra Care for Lips—Autumn Harp, 1

Home Shopping/Mail Order

Apricot Lip Balm—Body Shop, 2
Clear Lip Moisturizer—Finelle Cosmetics, 2
Herbal Moist Face Cream—Earth Gifts, 1
Jojoba Lip Balm—KSA Jojoba, 1
Key West Lip Balm—Key West Fragrance &
    Cosmetic Factory, Inc., 1
Key West Lip Balm Stick—Key West Fragrance &
    Cosmetic Factory, Inc., 1
Kiwi Fruit Lip Balm—Body Shop, 2
Lip Balm—The Face Food Shoppe, 1
Lip Savvy—Strong Skin Savvy, Inc., 2

Moisturizing Lip Primer—Oriflame International, 3
Morello Cherry Lip Balm—Body Shop, 2
Protective LipMender Gloss—Kallima International, Inc., 1
Shaklee Naturals Lip Conditioner—Shaklee U.S., Inc., 1

Specialty Stores

Aloe Vera Lip Balm—Scarborough and Company, 1
Jojoba Lip Balm—Scarborough and Company, 1
Lip Balm—Crabtree & Evelyn, Ltd., 1
Lip Balms—Body Shop (Calif.), 2

LIQUID BODY SOAPS

General Retail Stores

Aloe Vera Shower Gel—TerraNova, 1
Apricot Shower Gel—TerraNova, 1
Baby's Breath Baby Scent—North Country Soap, 1
Botanicals Glycerine Soap—Smith & Vandiver, Inc., 1
China Lily Shower Gel—TerraNova, 1
China Mist—TerraNova, 3
China Mist Shower Gel—TerraNova, 1
Citronella—North Country Soap, 1
De Maris Bath & Shower Gel—Cosmyl Cosmetics, 3
Earth Friendly Hand Soap—Venus Laboratories, Inc., 1
Fine Toiletries Glycerine Soap—Smith & Vandiver, Inc., 1
Fisherman 142—North Country Soap, 1
Fuzz—North Country Soap, 1
Golden Lotus Aloe Creme Face Soap—Mountain Fresh Products, 1

Golden Lotus Papaya Creme Bath Soap—Mountain Fresh Products, 1
Jergens Liquid Soap—Andrew Jergens, 3
Lady Slipper Woody Violet—North Country Soap, 1
Moon Shadows Shower Gel—TerraNova, 1
New Shower Gel—Botanicus, 1
100% Natural Liquid Glycerine Soap—Clearly Natural, 1
Paradise Mist Shower Gel—TerraNova, 1
Pikaki Shower Gel—TerraNova, 1
Pure Pleasure Liquid Soap—Surrey, Inc., 1
Rain Shower Gel—TerraNova, 1
Shabbos Soap—Adwe Laboratories, 1
Sinclair & Valentine Glycerine Soap—Smith & Vandiver, Inc., 1
Tropical Gardenia Shower Gel—TerraNova, 1

Hair Salons

Biokosma Cleopatra Shower Bath—Dr. Grandel, Inc., 2
Biokosma Dusch-X Shower Bath—Dr. Grandel, Inc., 1
Biokosma Sandalwood Shower Bath—Dr. Grandel, Inc., 2
Body Wash—Jelene International, 1
Instant Hand Sanitizer—Alpha 9, 1

Health Food Stores

Almond Moisture Soap—Kiss My Face, 1
Almond Oil Pure Castile Soap—Dr. Bronner, 1
Aloe Vera Body Wash—Rachel Perry, 1
Aloe Vera Facial Soap in a Jar—Jason Natural Products, 1
Aloe Vera Liquid Satin Soap—Jason Natural Products, 1
Aloe Vera Satin Body Wash—Jason Natural Products, 1
Aloe-Gel Lathering Hand Cleanser—Earth Science, 1

Olive & Aloe Moisture Soap—Kiss My Face, 1
Olive Oil Hand Soap—Heritage Store, 1
Oxyfresh Gel—Oxyfresh, 1
Peaches & Cream Moisture Soap—Kiss My Face, 1
Peppermint Oil Pure Castile Soap—Dr. Bronner, 1
Spearmint Leaf Body Scrub—Rachel Perry, 1
Spearmint Leaf Revitalizing Body Wash—Rachel
   Perry, 1
Swimmers & Sport Body Wash—Jason Natural
   Products, 1
Tea Tree Liquid Soap—Desert Essence, 1

Home Shopping/Mail Order

Active-Aloe Bath & Shower Suds—Kallima
   International, Inc., 1
Allens Liquid Soap with Aloe—Allens Naturally, 1
Aloe Body Gelee—Magic of Aloe, 3
Apache Gold Skin Gelee—Hobe Laboratories, Inc., 1
Bath & Shower Gelee—Wachter's, 1
Body Massage Wash—Key West Fragrance &
   Cosmetic Factory, Inc., 1
Body Satin—Granny's Old Fashioned Products, 1
Body Scrub—Oriflame International, 3
Body Shampoo—Nutri-Metrics International, Inc., 1
Body Smooth—Oriflame International, 3
Body Wash with Aloe—Key West Fragrance &
   Cosmetic Factory, Inc., 1
Cleansing Gel—La Costa Products International, 1
Cleansing Gel—Mary Kay Cosmetics, Inc., 3
Dewberry Body Shampoo—Body Shop, 2
Golden Glycerin Soap—Patricia Allison, 1
Herb Body Shampoo and Gel—Body Shop, 2
Jojoba Liquid Glycerin Soap—KSA Jojoba, 1
Jojoba Liquid Soap—KSA Jojoba, 1
L'Arome Echoes Body Shampoo—L'Arome USA,
   Inc., 1
Latis Natural Soap—Livos Plant Chemistry, 1
Meadow Blend Soap-Free Liquid Cleanser—Shaklee
   U.S., Inc., 1

Rice Bran Body Scrub—Body Shop, 2
Shower & Bath Gel—La Costa Products
 International, 1
Strawberry Body Shampoo and Gel—Body Shop, 2
Tea Rose Shower Gel—Body Shop, 2
Tropics Shower Gel—Body Shop, 2

Specialty Stores

Almond Bath Gel—Crabtree & Evelyn, Ltd., 1
Aloe Vera Bath Gel—Crabtree & Evelyn, Ltd., 1
Apricot Bath Gel—Crabtree & Evelyn, Ltd., 1
Avocado Bath Gel—Crabtree & Evelyn, Ltd., 1
Bath & Shower Herbal Gels—Scarborough and
 Company, 1
Birch Bath Gel—Crabtree & Evelyn, Ltd., 1
Birch Sauna Bath Gels—Crabtree & Evelyn, Ltd., 1
Body Shampoo—Body Shop (Calif.), 1
Buttermilk Bath Gel—Crabtree & Evelyn, Ltd., 2
Damask Rose Bath Gel—Crabtree & Evelyn, Ltd., 1
Eau de Cologne Bath Gel—Crabtree & Evelyn,
 Ltd., 1
Gardenia Bath Gel—Crabtree & Evelyn, Ltd., 1
Glycerine Facial Soap—Scarborough and
 Company, 1
Goatmilk Bath Gel—Crabtree & Evelyn, Ltd., 2
Jojoba Bath Gel—Crabtree & Evelyn, Ltd., 1
Lavender Bath Gel—Crabtree & Evelyn, Ltd., 1
Lily of the Valley Bath Gel—Crabtree & Evelyn,
 Ltd., 1
Lime Bath Gel—Crabtree & Evelyn, Ltd., 1
Liquid Soap—i Natural Skin Care and Cosmetics, 3
Millefleurs Bath Gel—Crabtree & Evelyn, Ltd., 1
Persian Lilac Bath Gel—Crabtree & Evelyn, Ltd., 1
Rain Body Shampoo—Body Shop (Calif.), 1
Rosewater Bath Gel—Crabtree & Evelyn, Ltd., 1
Sandalwood Bath Gel—Crabtree & Evelyn, Ltd., 1
Sea Aloe Body Shampoo—Body Shop (Calif.), 3
Shower/Bath Gel—Bare Escentuals/Dolphin
 Acquisition Corp., 2

Tiara Tahiti Bath Gel—Crabtree & Evelyn, Ltd., 1
Vetiver Bath Gel—Crabtree & Evelyn, Ltd., 1
Wheatgerm Bath Gel—Crabtree & Evelyn, Ltd., 1
White Ginger Bath Gel—Crabtree & Evelyn, Ltd., 1

M

MAKEUP REMOVERS

General Retail Stores

Ardell Eye Treats—Ardell International, 2
Ardell Non-Oily Eye Treats—Ardell International, 2
Delicate Eye Make-up Remover—Cosmyl
Cosmetics, 3
Eye Make-up Remover—Clientele, 1
Eye Q's—Andrea International Ind., 2
Gentle Eye Makeup Remover Gel—Color Me
Beautiful, 1
Instant Makeup Remover Pads—Color Me
Beautiful, 1
IsaDora Eye Make-up Remover—Cosmetic
Technology, Inc.,2
*Naturessence Jojoba & Avocado Oil Make-up
Remover*—Avanza Corp.,1
Quick 'n' Gentle—Ardell International, 2
*Talika Eye Make-up Remover for Waterproof
Mascara*—Jean Pax, Inc., 1
Talika Eye Make-up Remover—Jean Pax, Inc., 1
Talika Gel Eye Make-up Remover—Jean Pax, Inc., 1

Hair Salons

Eye Makeup Remover—Studio Magic, 2
Make-up Remover—Jelene International, 1

Health Food Stores

Cleansing Milk & Makeup Remover—Reviva Labs, 2
Evening Primrose Oil Eye Makeup Remover—Rachel
   Perry, 2
Eye Make-up Removing Lotion—Beauty Without
   Cruelty, Ltd., 1
Eye Makeup Remover—Paul Penders, 1
Gentle Chamomile Eye Makeup Remover—Natural
   Solutions, 1
Makeup Remover—Borlind of Germany, 1
Nature Jojoba & Olive Oil Make-up Remover—
   Avanza Corp., 1
Nature's Gentle Eye Make-up Remover Lotion—
   Avanza Corp., 1
Nature's Waterproof Eye Make Up Remover Liquid—
   Avanza Corp., 1

Home Shopping/Mail Order

Chamomile Eye Make-up Remover—Body Shop, 2
Chamomile Eye Makeup Remover—Michel
   Constantini NaturalCosmetics, 2
Eye Makeup Remover—Mary Kay Cosmetics, Inc., 3
Eye Makeup Remover Disques—Finelle Cosmetics, 2
Gel Eye Makeup Remover—Magic of Aloe, 3
Lipstick & Eye Makeup Remover—Jacklyn Cares, 2
Lynndi's Face Oil—The Face Food Shoppe, 1
Make-Up Remover Pads—Patricia Allison, 1
No Rub Makeup Remover—Irma Shorell, 2

Specialty Stores

Rosemary Eye Makeup Remover—i Natural
   Skin Care and Cosmetics, 3
Rosemary Eye Makeup Remover Pads—i Natural
   Skin Care and Cosmetics, 3

MASSAGE OILS

General Retail Stores

China Lily Luxury Oil—TerraNova, 1
China Mist Luxury Oil—TerraNova, 1
China Rose Luxury Oil—TerraNova, 1
Jacob Hooy Massage Oils—Baudelaire, 1
Moon Shadows Luxury Oil—TerraNova, 1
Pikaki Luxury Oil—TerraNova, 1
Rain Luxury Oil—TerraNova, 1
Sweet Almond Oil—TerraNova, 1
Vanilla Creme Lotion—TerraNova, 1

Hair Salons

Aromatherapy Body Massage Oils—Jurlique D'Namis, Ltd., 1
Biokosma Massage Oils—Dr. Grandel, Inc., 1
Biokosma Sea Line Massage Cream—Dr. Grandel, Inc., 1
Biokosma Seaweed Massage Cream—Dr. Grandel, Inc., 1
Silhouette Massage Oil—Dr. Babor Natural Cosmetics, 2

Health Food Stores

Arnica Massage Oil—Weleda, Inc., 1
Aromatherapy Massage and Bath Oils—Aura Cacia, 1
Balance Body Oil—Desert Essence, 1
Egyptian Oil—Heritage Store, 1
Energy Body Oil—Desert Essence, 1
Exotic Massage Oils—Liberty Natural Products, 1
French Massage Formula—Rachel Perry, 1
Gold Rush—Heritage Store, 1
Golden Magic—Heritage Store, 1
Green Herbal Massage Gel—Natural Solutions, 3
Jacki's Magic Lotion—Jacki's Magic Lotion, 2

Light Touch/Light Mountain Massage Oils—Lotus Light, 1

Love Body Oil—Desert Essence, 1

Love Butter—Aura Cacia, 1

Massage Cremes—Jason Natural Products, 1

Massage Oil—Michael's Health Products, 1

Massage Oils—Aroma Vera, 1

Massage Oils—Jason Natural Products, 1

Massage Oils—Liberty Natural Products, 1

Massage Oils—Heritage Store, 1

Rejuvenating Herbal Massage Oil—Lakon Herbals, 1

Relaxation Body Oil—Desert Essence, 1

Seaherbal Massage Lotion—Aubrey Organics, 1

Selena Healing Massage Oil—WiseWays Herbals, 1

Seven Golden Massage Oils—Natural Solutions, 1

Soothing Touch Massage Lotion—Sunshine Products Group, 1

Soothing Touch Massage Oils—Sunshine Products Group, 1

Sports Massage Oil—Rainbow Research Corp., 2

Swasthya Ayurvedic Massage Oils—Auromere, Inc., 1

Tea Tree Oil—Desert Essence, 1

Home Shopping/Mail Order

After Massage Toner—Key West Fragrance & Cosmetic Factory, Inc., 1

Aloe Massage Lotion—Grace Cosmetics/Pro-Ma Systems, 1

Arnica-Chamomile Massage & Bath Therapy—Indian Creek, 1

Body Massage Oil—Body Shop, 2

Glowing Touch Revitalizing Skin Care Oils—G. T. International, 1

Massage Essence—Wachter's, 1

Massage Lotion—Key West Fragrance & Cosmetic Factory, Inc., 1

Massage Oils—Earth Gifts, 1

Massage Oils—Jeanne Rose, 2
Multi-Active Hot & Cold Massage Lotion—Kallima
   International, Inc., 1
Ready Mixed Massage Oils—Shirley Price
   Aromatherapy, 1
Rich Massage Lotion—Body Shop, 2
Soothing Touch Skin Care Oil—G. T.
   International, 1
Spa Massage Oil—La Costa Products
   International, 1
Stimulating Facial Massage Oil—G. T.
   International, 1
Vitacore Conditioning Massage Therapy—La Costa
   Products International, 1
Vitamin Massage Oil—Patricia Allison, 1

Specialty Stores _____

Amaretto Massage Balm—Body Shop (Calif.), 1
5-Oil Blend—Bare Escentuals/Dolphin Acquisition
   Corp., 2
Honeydew Massage Balm—Body Shop (Calif.), 3
Kama Sutra—Potions & Lotions, 3
Massage Lotion—Bare Escentuals/Dolphin
   Acquisition Corp., 2
Massage Lotion—Body Shop (Calif.), 3
Massage Lotion—Potions & Lotions, 3
Massage Oil—Body Shop (Calif.), 1
Massage Oil—Potions & Lotions, 3
Strawberry Massage Balm—Body Shop (Calif.), 3
Sweet Almond Oil—Body Shop (Calif.), 1

MEN'S HAIR CARE PRODUCTS

General Retail Stores _____

Clubman Hair Tonic—A.I.I. Clubman, 1

Clubman Liquid Hair Spray—A.I.I. Clubman, 1
Clubman Styling Gel—A.I.I. Clubman, 1
Collagen Balsam—A.I.I. Clubman, 3
Grooming and Styling Spray—A.I.I. Clubman, 1
Supreme Hair Spray—A.I.I. Clubman, 1
Supreme Non-Aerosol Hair Spray—A.I.I.
  Clubman, 1

Home Shopping/Mail Order _____

Hairgel—Gruene, 1
Slick Hair Styler—Body Shop, 2

MEN'S POWDER AND TALC

General Retail Stores _____

Clubman Talc—A.I.I. Clubman, 2

Home Shopping/Mail Order _____

Talc No. 1 Samarkand—Body Shop, 2

Specialty Stores _____

Men's Range Talcum Powder—Crabtree &
  Evelyn, Ltd., 1

MEN'S SHAMPOOS AND CONDITIONERS

General Retail Stores _____

Aloe Vera Grooming Shampoo—TerraNova, 3
Country Club Shampoo—A.I.I. Clubman, 1

Home Shopping/Mail Order —————————————

Natural Revitalizing Conditioner—Gruene, 1
Natural Revitalizing Shampoo—Gruene, 1
Rhassoul Mud Shampoo—Body Shop, 2

Specialty Stores —————————————

Hair Conditioner—Crabtree & Evelyn, Ltd., 1
Protein Shampoo—Crabtree & Evelyn, Ltd., 1

MEN'S SKIN LOTIONS

Health Food Stores —————————————

Daily Workout Sesame Moisture Lotion—Natural
     Solutions, 3
Face and Body Skin Care Balm—Earth Science, 1
Skin Conditioner Lotion for Men—Reviva Labs, 2

Home Shopping/Mail Order —————————————

Aloe Body Lotion—Gruene, 1
Aloe Moisture for Men—Grace Cosmetics/Pro-Ma
     Systems, 1
Gruene Moisture Formula—Gruene, 1
Mr. Aloe Moisture Cream—Magic of Aloe, 3
Skin Management Conditioner—Mary Kay
     Cosmetics, Inc., 3
Tioga Men's Skin Conditioner—Shaklee U.S., Inc., 1

MEN'S SKIN TONERS

Hair Salons —————————————

Babor Homme Face Vitalizer—Dr. Babor
     Natural Cosmetics, 1

Home Shopping/Mail Order

97% Aloe Skin Toner—Grace Cosmetics/Pro-Ma
   Systems, 1
Oil Absorber—Mary Kay Cosmetics, Inc., 3
Toner—Mary Kay Cosmetics, Inc., 3

MEN'S SOAPS

General Retail Stores

Aloe Vera Shower Gel—TerraNova, 1
Desert Rain Glycerine Soap—TerraNova, 1

Health Food Stores

Daily Workout Avocado Face Scrub—Natural
   Solutions, 1
Daily Workout Herbal Body Soak—Natural
   Solutions, 1
Daily Workout Soap—Natural Solutions, 3

Home Shopping/Mail Order

Aloe Deep Cleanser—Grace Cosmetics/Pro-Ma
   Systems, 1
Body Washing—Gruene, 1
Cleansing Bar—Mary Kay Cosmetics, Inc., 3
Colloidal Cleanser—Biogime, 1
Daily Cleansing Scrub—Gruene, 1
Face Protector—Body Shop, 2
Face Scrub—Body Shop, 2
Face Wash—Body Shop, 2
Foaming Cleanser—Biogime, 1
Mr. Aloe Washing Cleanser—Magic of Aloe, 3
Rhassoul Mud Soap—Body Shop, 2

Specialty Stores

Men's Range Soap—Crabtree & Evelyn, Ltd., 1

METAL CLEANERS AND POLISHES

General Retail Stores

Copper Glo—SerVaas Laboratories, 1
Liquid Copper Glo—SerVaas Laboratories, 1
Magic Silver Dip—Magic American Corp., 1
Stainless Steel Magic—Magic American Corp., 1

MILDEW CONTROL PRODUCTS

General Retail Stores

Magic Mildew Stain Remover—Magic
American Corp., 1

Home Shopping/Mail Order

X158 Mildew Control—AFM Enterprises, Inc., 1

MOISTURIZERS

General Retail Stores

Aloe & E Moisturizing Cream—Colonial Dames
Co., Ltd., 1
Aloe Vera Body Moisturizer—TerraNova, 3
Ambi Skin Moisturizing Creme—Kiwi Brands,
Inc., 3
Apiana Honey Moisturizing Cream—Baudelaire, 2
Botanical Moisture Fluid—Smith & Vandiver,
Inc., 1

Botanical Moisture Replenisher—Smith & Vandiver, Inc., 1

Cucumber Moisture-Rich Cream Cleanser—TerraNova, 3

Daily Vitamins Moisture Cream—TerraNova, 1

Daytime Moisture Concentrate—Clientele, 1

Extra Protective Hydrating Complex—Color Me Beautiful, 3

Fluid Formula Moisturizer—Colonial Dames Co., Ltd., 1

Fragylis Hydradouce Moisturizer—Jean Pax, Inc., 3

Gentle Hydrating Lotion—Cosmyl Cosmetics, 3

Hydrating Tonic—Color Me Beautiful, 1

Jacob Hooy Skin Creams—Baudelaire, 2

Light Moisturizer—Color Me Beautiful, 2

Lightweight Moisture Complex—Color Me Beautiful, 1

Moisture Concentrate—Clientele, 1

Moisturizing Skin Balancer—Cosmyl Cosmetics, 3

Natural Touch Moisturizer—Natural Touch, 3

Naturessence Aloe Vera & PABA All Day Moisturizer—Avanza Corp., 1

Naturessence NaPCA & PABA Daytime Facial Moisturizer—Avanza Corp., 1

9-5 Light Soothing Aloe Vera Moisture Creme—TerraNova, 2

Oil Free Facial Moisturizer—Botanicus, 1

Oil Free Regulating Fluid—Color Me Beautiful, 1

Plus IV Deep Moisturizer—Tyra Skin Care, 2

Premiere Moisturizer—Tyra Skin Care, 2

Premiere Normalizer—Tyra Skin Care, 2

Rich Moisturizer—Color Me Beautiful, 2

Vitamin E Fluid Moisturizer—Colonial Dames Co., Ltd., 1

Vitamins A, D & E Body Moisturizer—TerraNova, 3

Wysong Dermal—Wysong Corp., 1

Hair Salons

Hydrate—Mastey de Paris, 1

Hydrovita Day Cream—Dr. Babor Natural
  Cosmetics, 1
Miraculous Beauty Replenisher—Aveda Corp., 1
Moiste—Mastey de Paris, 1
Moisturee—Mastey de Paris, 1
Nutre-Moist21 Skin Nutrient Moisture—Focus 21, 1
Protective Creme—Jelene International, 2
Super Naturals Aloe Vera Moisturizing Cream—
  North Country Naturals, 2
Super Naturals Elastin Moisturizing Cream—North
  Country Naturals, 3
Super Naturals Vitamin E Moisturizing Cream—
  North Country Naturals, 2
Very Nourishing Moisturizing Cream—Aveda
  Corp., 1

Health Food Stores

Active Moisture Formula—Orjene Natural
  Cosmetics, 3
ADE Cream—Carlson Laboratories, 3
ADE Ointment—Carlson Laboratories, 3
Almond Aloe Light & Silky Moisturizer—Earth
  Science, 1
Almond Cream—Alexandra Avery, 1
Aloe and E Moisture Creme—Beauty Without
  Cruelty, Ltd., 1
Aloe Vera Deep Skin Moisturizer—Orjene
  Natural Cosmetics, 3
Aloe Vera Lavender Moisturizer—Paul Penders, 1
Aloe Vera Moisturizer—Orjene Natural Cosmetics, 2
Aloegen Hydrating Body Emulsion—Levlad, 3
Aloegen Hydrating Day Emulsion—Levlad, 3
Aloe-Herb Maximum Moisture Cream—Natural
  Bodycare, 1
American Ginseng Extra Rich Moisturizer—Natural
  Solutions, 3
AQUA-A Daytime Moisturizer—Jason Natural
  Products, 1

Iris Moisturizing Cream—Weleda, Inc., 2

Jojoba Aloe Vera Moisture Cream—Desert Essence, 1

Key-E Cream—Carlson Laboratories, 3

Key-E Ointment—Carlson Laboratories, 1

Key-E Spray—Carlson Laboratories, 1

Lecithin-Aloe Moisture Retention Cream—Rachel Perry, 1

Lettuce Moisturizer Lotion—Reviva Labs, 2

Light Touch Coconut Oil Moisturizer—Lotus Light, 1

Lily Herbal Moisturizer—Lily of Colorado, 1

Lily Moisturizing Cream—Lily of Colorado, 1

LL Moisturizing Ampoules—Borlind of Germany, 1

Maintenance for Young Skin—Aubrey Organics, 1

Mandarin Magic Moisturizer—Aubrey Organics, 1

Moisture Glow with Color Tone—Reviva Labs, 2

NaPCA & Aloe Moisturizer—Reviva Labs, 2

NaPCA & PABA Daytime Facial Moisturizer—Avanza Corp., 1

Nature Aloe Vera & PABA All-Day Moisturizer—Avanza Corp., 1

Nature Cocoa Butter & Vitamin E Skin Moisturizer—Avanza Corp., 1

Nature's Active Nutrisome & Vitamin E Moisture Lotion—Avanza Corp., 1

Nature's Gate Moisture Cream—Levlad, 1

Normalizing Lotion w/Hyaluronidase—Reviva Labs, 3

Oil Free Hydrator—Beauty Without Cruelty, Ltd., 1

Oil Free Moisture Creme—Beauty Without Cruelty, Ltd., 1

Oil-Free Moisturizer With NaPCA—Kiss My Face, 1

Oily Skin Moisturizer Cream—Reviva Labs, 3

Olive & Aloe Moisturizer—Kiss My Face, 1

Orchid Oil Moisturizer—Orjene Natural Cosmetics, 1

Papaya Enzyme Moisturizer Creme—Orjene Natural Cosmetics, 1

Peach Day Cream—Borlind of Germany, 1

Winter Wheat Moisturizer—Natural Solutions, 3
Z Herbal Cream—Borlind of Germany, 1
Z Moisturizing Ampoules—Borlind of Germany, 2
ZZ Herbal Day Cream—Borlind of Germany, 1

Home Shopping/Mail Order

All-Day Moisturizing Cream—Sombra, 1
Aloe Crystal Gel Concentrate—Sombra, 1
Aloe Crystal Gel Light—Sombra, 1
Aloe Crystal Gel Regular—Sombra, 1
Aloe Moist Dew w/Jojoba—Grace Cosmetics/Pro-Ma
    Systems, 1
Aloe Moist—Nutri-Metrics International, Inc., 1
Aloe Moisture Cream—Barbizon International, 3
Aloe Moisture Cream—Grace Cosmetics/Pro-Ma
    Systems, 1
Aloe Moisture Lotion—Jacklyn Cares, 2
Aloe Perfection—Magic of Aloe, 3
Aloe Secrets—Nutri-Metrics International, Inc., 1
Aloe Vera Moisture Cream—Body Shop, 2
Avocado Cream Moisturizer—Shaklee U.S., Inc., 1
Balancing Moisturizer—Mary Kay Cosmetics,
    Inc., 3
Body Milk—The Face Food Shoppe, 1
Carrot Moisture Cream—Body Shop, 2
Chamomile Moisturizing Complex—Indian Creek, 1
Contour/35—Irma Shorell, 2
Crystal Energy Hydrating Treatment—Ecco Bella, 1
Daily Revival—Avon Products Inc., 3
Day Moist—Key West Fragrance & Cosmetic
    Factory, Inc., 1
Desert Mist Moisturizer—Hobe Laboratories, Inc., 1
80 Plus w/Biotin—Magic of Aloe, 3
Elastin Collagen Lotion—Michel Constantini
    Natural Cosmetics, 3
Enriched Moisturizer—Mary Kay Cosmetics, Inc., 3
Extra Emollient Moisturizer—Mary Kay Cosmetics,
    Inc., 3

Extra Moisturizer—Shaklee U.S., Inc., 1
Herbal Moist Face Cream—Earth Gifts, 1
Herbal Moisture Lotion—Barbizon International, 3
Hydra-Complexe—Finelle Cosmetics, 2
Hydrating Geranium Creme—Wachter's, 1
Ida Grae Earth Venus Moisturizer—Nature's Colors, 2
Intensive Treatment Creme—Michel Constantini Natural Cosmetics, 3
Jojoba Moisture Cream—Body Shop, 2
Light Moisturizer—Shaklee U.S., Inc., 1
Light Moisturizing Lotion—Kimberly Sayer, Inc., 1
Moisan 5X Moisturizer—Nutri-Metrics International, Inc., 1
Moisture Cream with Vitaimin E—Body Shop, 2
Moisture Cream—Shaklee U.S., Inc., 1
Moisture Cream—Shirley Price Aromatherapy, 1
Moisture Lotion—Magic of Aloe, 3
Moisture Lotion—Shaklee U.S., Inc., 1
Moisture Lotion—Shirley Price Aromatherapy, 1
Moisture Perfection—Oriflame International, 3
Moisture Protection—Oriflame International, 3
Moisture Renewal Treatment Cream—Mary Kay Cosmetics, Inc., 3
Moisture Shield—Avon Products, Inc., 1
Moisture/35—Irma Shorell, 2
Moisturizer—Finelle Cosmetics, 2
Moisturizing Creme with Collagen, Vitamin E, Aloe Vera &Panthenol—Bronson Pharmaceuticals, 3
Moisturizing Creme—La Costa Products International, 1
Moisturizing Gel—La Costa Products International, 1
Nutri-Moist—Nutri-Metrics International, Inc., 1
Oasis Moisture Balm—Patricia Allison, 1
Oil Control Lotion—Mary Kay Cosmetics, Inc., 3
Oily Moisture Lotion—Magic of Aloe, 3
Orchid Oil Moisturizer—Jacklyn Cares, 1

Protective Full Spectrum Rich Creme—Kallima
    International, Inc., 1
Protective Moisture Bloom Creme—Kallima
    International, Inc., 1
Protective Moisture Mist Lotion—Kallima
    International, Inc., 1
Protective Moisturizer Lotion—Kimberly Sayer,
    Inc., 1
Protective Moisturizer SPF 6—Finelle Cosmetics, 2
Proto-Collagen Moisturizer—Magic of Aloe, 3
Rich Moisture/Vita Moist—Avon Products, Inc., 3
Sheer Moisture—Irma Shorell, 2
Silken Day Moisture Creme—G. T. International, 1
Swedish Pollenique Active Day Moisture Lotion—
    Cernitin America, Inc., 1
Swedish Pollenique Balancing Moisture Formula—
    Cernitin America, Inc., 1
Swedish Pollenique Body Silk Moisturizer—
    Cernitin America, Inc., 1
Swedish Pollenique Pollen Gold Moisturizer—
    Cernitin America, Inc., 1
Ultra-Rich—Patricia Allison, 1
Under Makeup Moisturizer—Shaklee U.S., Inc., 1
VitaBalm—Patricia Allison, 1

Medical Professionals

Cielo Sunscreen/Moisture Lotion—Professional &
    Technical Services, Inc., 2
Pro-Gest Moisturizing Cream—Professional &
    Technical Services, Inc., 2

Specialty Stores

AD & E Moisturizer—Body Shop (Calif.), 3
Aloe Vera Moisturizer—Body Shop (Calif.), 3
Apricot Moist—i Natural Skin Care and
    Cosmetics, 3
Bio-Moisture—i Natural Skin Care and Cosmetics, 2

Citrus Day Skin Care—Potions & Lotions, 3
Collagen-Elastin: 14-1 Creme—Potions & Lotions, 3
Daytime Moisturizer—Potions & Lotions, 3
E.C.P. Maxima Creme—Potions & Lotions, 3
Extra Moist—i Natural Skin Care and Cosmetics, 3
Jojoba Oil Moisturizer—Body Shop (Calif.), 3
Light Lemon Moist—i Natural Skin Care and
   Cosmetics, 3
Milk of Aloe Moist—i Natural Skin Care and
   Cosmetics, 3
Moisture Gel—Body Shop (Calif.), 2
Natural Protection Tinted Day Cream—i Natural
   Skin Care and Cosmetics, 3
Sea Aloe Moisturizer—Body Shop (Calif.), 3
13 Super Cream—i Natural Skin Care and
   Cosmetics, 3
Vitamin Cream—i Natural Skin Care and
   Cosmetics, 3

MOTH REPELLENTS

Home Shopping/Mail Order

Cedar Herbal Sachet—Lady of the Lake Company, 1
Moth Away Sachet—Rathdowney, Ltd., 1

MOUTHWASHES

General Retail Stores

Astringent Mouthwash—Adwe Laboratories, 1
Tom's of Maine Natural Mouthwash—Tom's of
   Maine, 1

Dentists

Potentianted Doublemint Mouthwash—Oxyfresh, 1
Potentiated Natural Mouthwash—Oxyfresh, 1

Health Food Stores

Ipsadent Herbal Mouthwash—Heritage Store, 1
Mer-flu-an Mouthwash Concentrate—American
    Merfluan, Inc., 1
Mouthwash Concentrate—Weleda, Inc., 1
Natural Mint Mouthwash & Gargle—Aubrey
    Organics, 1
Tea Tree Oil Mouthwash—Desert Essence, 1
TIB Breath Fresheners—Liberty Products Corp., 1

MUSCLE RUBS AND LINIMENTS

Health Food Stores

Aussie Gold Icy Mineral Gel—Jason Natural
    Products, 1
Deep Muscle Therapy Sports Massage Oil—Lakon
    Herbals, 1
Muscle Treat Liniment—Heritage Store, 1
Roll A Rub—Breezy Balms, 2

Home Shopping/Mail Order

Great Stuff for Aches and Pains—Hobe
    Laboratories, Inc., 1

N

NAIL POLISH

General Retail Stores

All nail polishes—Cosyml Cosmetics, 3
IsaDora Cosmetique—Cosmetic Technology Inc., 2
Naturessence Nail Color—Avanza Corp., 1

Health Food Stores

All nail polishes—Borlind of Germany, 1
All nail polishes—Eva Jon Cosmetics, 3
Nail Colors—Beauty Without Cruelty, Ltd., 1
Nature Nail Polish—Avanza Corp., 1
Naturessence Nail Polish—Avanza Corp., 1

Home Shopping/Mail Order

All nail polishes—Magic of Aloe, 3
All nail polishes—Mary Kay Cosmetics, 3
Grace Cosmetics—Grace Cosmetics/Pro-Ma
    Systems, 1

NAIL POLISH DRYING AGENTS

General Retail Stores

Andrea Super Dry—Andrea International Ind., 2

DeLore Organic Nail Hardener & Polish Dryer—
DeLore Industries, 2
*Fast Touch Quick Dry—*Cosmyl Cosmetics, 3

Health Food Stores

*Professional Nail Dryer—*Ardell International, 2

Home Shopping/Mail Order

*Quick Dry—*Mary Kay Cosmetics, Inc., 3

NAIL POLISH REMOVERS

General Retail Stores

*Acetone Polish Remover—*Super Nail, 2
*Andrea Nail Polish Remover Pads—*Andrea
International Ind., 2
*Ardell Nail Polish Remover—*Ardell International, 2
*Ardell Polish Remover Gel—*Ardell International, 2
*Extra-Gentle Polish Remover—*Cosmyl Cosmetics, 3
IsaDora Nail Polish Remover— Cosmetic
Technology, Inc., 2
*Non-Acetone Mild Formula Polish Remover—*Ardell
International, 2
*Non-Acetone Polish Remover—*Super Nail, 2
*Polish Corrector Pen—*Ardell International, 2
*Polish Remover—*Adwe Laboratories, 1
*Pure Acetone Polish Remover—*Super Nail, 2
*Smearproof Polish Remover—*Cosmyl Cosmetics, 3
*Swirl Off—*Andrea International Ind., 2
*Unique Polish Remover—*DeLore Nails, 2

Health Food Stores

*Nail Polish Remover—*Borlind of Germany, 1

Home Shopping/Mail Order _____

Advanced Nail Color Remover—Mary Kay
Cosmetics, 3

NIGHT CREAMS

General Retail Stores _____

Extra Rich Night Cream—Colonial Dames Co.,
Ltd., 1
Overnight Collagen Treatment Moisture Cream—
TerraNova, 3
Regulating Night Therapy—Color Me Beautiful, 1

Hair Salons _____

Biokosma Night Cream—Dr. Grandel, Inc., 1
Night Nourishing Creme—Jelene International, 1
Super Naturals Aloe Vera Intensive Night Cream—
North Country Naturals, 2

Health Food Stores _____

Aloegen Overnight Renewal Emulsion—Levlad, 3
Aloe-Herb Night Renewal Cream—Natural
Bodycare, 1
Apricot Age-Controlling Night Creme—Earth
Science, 1
Apricot-Night Cream—Earth Science, 1
Avocado Ginseng Night Cream—Paul Penders, 1
Blackmores Avocado Night Creme Plus—Solgar
Vitamin Co., 1
Blackmores White Lily Hydrating Night Creme—
Solgar Vitamin Co., 1
Celltherapy Cellular Night Repair—Aubrey
Organics, 3

Collagen Night Time Moisture Renewel Creme—
Orjene Natural Cosmetics, 3
Dream Creme—Alexandra Avery, 1
Elastin Night Cream—Reviva Labs, 3
Eva Jon Night Creme—Eva Jon Cosmetics, 3
Extra-Rich Night Cream—Reviva Labs, 3
F Night Cream—Borlind of Germany, 1
FF Rose Dew Cream—Borlind of Germany, 1
Intensive Night Care—Color Me Beautiful, 1
InterCell Night Gel—Reviva Labs, 3
Iris Night Cream—Weleda, Inc., 2
LL Bi-Aktiv Liposome Emulsion—Borlind of
Germany, 1
LL Regeneration Night Creme—Borlind of
Germany, 2
N Cream Supreme—Borlind of Germany, 1
Oily Skin Night Cream—Reviva Labs, 2
Rose Dew Night Cream—Borlind of Germany, 1
*Truly Moist Extra Dry NaPCA Night Facial
Treatment*—Desert Naturels, 1
*Truly Moist Original Formula NaPCA Night Facial
Treatment*—Desert Naturels, 1
U Herbal Night Cream—Borlind of Germany, 1
Wheatgerm Honeysuckle Night Cream—Paul
Penders, 1
Z Herbal Cream for Night—Borlind of Germany, 2
ZZ Herbal Night Cream—Borlind of Germany, 1

Home Shopping/Mail Order

Aloe Night Nurture Cream—GraceCosmetics/
Pro-Ma Systems, 1
Extra Emollient Night Cream—Mary Kay
Cosmetics, 1
Moisture Night Creme—Magic of Aloe, 3
Night Cream—Shirley Price Aromatherapy, 1
Night Replenishment—Oriflame International, 3
Nightmoist—Key West Fragrance & Cosmetic
Factory, Inc., 1

Nighttime Recovery System—Mary Kay
    Cosmetics, 3
Nite Caress Night Creme—Wachter's, 1
Rich Night Cream with Vitamin E—Body Shop, 2
So Rare Mist Night Creme—Nutri-Metrics
    International, 2
Vita-Magic Night Creme—Patricia Allison, 1
Vital Night Cream—Kimberly Sayer, Inc., 1
Vitamin A, D, E & B5 Night Cream—Sombra, 1

Specialty Stores ───────────────────

Bee Pollen Night Cream—Potions & Lotions, 3
Extra Rich Night Cream—Body Shop (Calif.), 3

OFFICE MISTAKE CORRECTORS

General Retail Stores ───────────────────

Boo Boo Goo Correction Fluid—International
    Rotex, 1
Coverit—Wite-Out Products, 1
Fax-Out—Wite-Out Products, 1
Non-Flaking Typing Correction Film—Wite-Out
    Products, 1
Pen-FC—Wite-Out Products, 1
Pen-TH—Wite-Out Products, 1

Rotex Bond White Correction Fluid—International
    Rotex, 1
Rotex Correction Pen for Typed Errors—
    International Rotex, 1
Rotex Correction Pen for Written Errors—
    International Rotex, 1
Rotex Correctsafe Correction Fluid—International
    Rotex, 1
Rotex for Copies Only—International Rotex, 1
Rotex for Ink—International Rotex, 1
Rotex Thinner—International Rotex, 1
Rotex Typing Correction Film—International
    Rotex, 1
Vaporsafe—Wite-Out Products, 1
Wite-Out for Copies—Wite-Out Products, 1
Wite-Out For Everything—Wite Out Products, 1
Wite-Out for Typewriters—Wite-Out Products, 1
Wite-Out Natural Dry—Wite-Out Products, 1
Wite-Out Thinner—Wite-Out Products, 1

OFFICE SUPPLIES

General Retail Stores

Big Boss Instant Signs—International Rotex, 1
D-inky Hand Cleaning Pads—Wite-Out Products, 1
Embossing Label Tape—International Rotex, 1
Label Makers—International Rotex, 1
Packing List Envelopes—International Rotex, 1
Photo Copier Labels—International Rotex, 1
Pinky Tac—Wite-Out Products, 3
Rotex Replacement Ribbons and Correcting Tapes—
    International Rotex, 1
Security Pens and Refills—International Rotex, 1
1300 Tape Printer & Supplies—International
    Rotex, 1

ORAL CARE PRODUCTS

Health Food Stores _____

Ipsab Herbal Gum Treatment—Heritage House, 1
Peri-Dent Herbal Gum Massage—Home Health
    Products, 2

PAIN RELIEF PRODUCTS

Health Food Stores _____

AhhhLoe Ice for Pain Relief—Reviva Labs, 2
Pain-Aid—Michael's Health Products, 2
Soothe Pain Relief Lotion—Home Health Products, 2

Home Shopping/Mail Order _____

Comfortcaine—Key West Fragrance & Cosmetic
    Factory, Inc. 1
Triflora Analgesic Gel—G. T. International, 1
Yi'ang Analgesic Balm—Wachter's, 1

PAINT

Home Shopping/Mail Order _____

Aidu-Radiator Paint—Livos Plant Chemistry, 1

Albion-White Wash Paint—Livos Plant Chemistry, 1
Amellos-Solvent Free Paint—Livos Plant
    Chemistry, 1
Canto-Satin Enamel Paint—Livos Plant
    Chemistry, 1
CEM Bond Concrete and Masonry Paint—AFM
    Enterprises, 1
Dubron-Natural Resin Wall Paint—Livos Plant
    Chemistry, 1
Safecoat All Purpose Enamel—AFM Enterprises, 1
Safecoat Enamel Water Base Semi-Gloss—AFM
    Enterprises, 1
Safecoat Paint—AFM Enterprises, 1
URA-Earth Colors—Livos Plant Chemistry, 1
Vindo-Enamel Paint—Livos Plant Chemistry, 1

PAINT AND
VARNISH STRIPPERS

Home Shopping/Mail Order

M P Stripper—AFM Enterprises, Inc., 1
Stripper 66—AFM Enterprises, Inc., 1

PERMANENTS
AND BODY WAVES

General Retail Stores

My Skin Permanents—Dena Corp., 1

Hair Salons

Aromatherapy Permanent Waving System—Aveda
    Corp, 1
Permanent Waves—Focus 21, 1
Revise—Paul Mazzotta, Inc., 1
SeaPlasma—Focus 21, 1
Self-Timing Acid Perm—Focus 21, 1
Sine Wave—JOICO Laboratories, 1
Solar Wave—JOICO Laboratories, 1
Sukesha Perm Systems—Chuckles, Inc., 1
Tetra S.T.A.R. Perm—JOICO Laboratories, 1

PET FIRST AID PRODUCTS

General Retail Stores

Hot Spot Oil—Tender Loving Care Pet Products, 1
Scale and Fungus Remover for Birds—Tender
    Loving Care Pet Products, 1

Health Food Stores

Royal Herbal Skin Ointment—Pet Connection, 1

PET FOODS AND SUPPLEMENTS

General Retail Stores

Lifesource Health Treats—Wysong Corp., 3
Lifesource Premium Pet Foods—Wysong Corp., 3
Maxi-Tone Pet Food Supplement—Tender Loving
    Care Pet Products, 1
PetGuard Premium Canned Food for Cats—
    PetGuard, 3
PetGuard Premium Canned Food for Dogs—
    PetGuard, 3

PetGuard Premium Dry Dog and Cat Food—
PetGuard, 3
*Wysong Canine and Feline Diets—*Wysong Corp., 3

Health Food Stores _____

*VitaL-21-Plus Vitamins—*Pet Connection, 3

PET GROOMING AIDS

General Retail Stores _____

*Cream Rinse—*Tender Loving Care Pet Products, 1

Health Food Stores _____

*All-Natural Cream Coat Conditioner—*Natural
Bodycare, 1
*Herbal Pet Coat Conditioner —*Nature's Gate, 1
*My Pet Herbal Pet Coat Conditioner —*Levlad, 1
*Organimals Dip & Creme Rinse—*Aubrey
Organics, 1
*Organimals Organic Grooming Spray—*Aubrey
Organics, 1

Home Shopping/Mail Order _____

*Jojoba Oil for Pets—*KSA Jojoba, 1

PET ODOR AND STAIN REMOVERS

General Retail Stores _____

*Citressence Pet Spray—*Wysong Corp., 1

Dog Body Deodorizer & Cleaner—Pets 'N People, 1
Nature's Miracle Cage & Aviary Deodorizer—Pets 'N People, 1
Nature's Miracle Litter Deodorant—Pets 'N People, 1
Nature's Miracle Skunk Odor Remover—Pets 'N People, 1
Nature's Miracle Stain & Odor Remover—Pets 'N People, 1
Pet Mess Easy Clean-up—Pets 'N People, 1
Stable, Tack & Riding Apparel Cleaner—Pets 'N People, 1

Health Food Stores

OutSTAINing Enzymatic Stain & Odor Remover—Pet Connection, 1
Pet Air—Mia Rose Products, 1
Tough Job Heavy-Duty Cleanser—Natural Bodycare, 1

PET PRODUCTS—MISCELLANEOUS

General Retail Stores

Wysong Litter Lite—Wysong Corp., 1

Health Food Stores

Saturday Nite Organic Catnip—Pet Connection, 1

Home Shopping/Mail Order

Catnip Toy—Rathdowney, Ltd., 1
Not Here Doggy!—Rathdowney, Ltd., 1
Not Here Kitty!—Rathdowney, Ltd., 1

PET SHAMPOOS

General Retail Stores

Shampoo—Tender Loving Care Pet Products, 1
Winner's Circle for Pets—North Country Soap, 1
Woof Glycerine Soap—Abbaco, Inc., 1

Health Food Stores

All-Natural Cleanser—Natural Bodycare, 1
All-Natural Shampoo with Herbs—Natural
   Bodycare, 1
Citrus Supreme Shampoo—Xavier, 1
My Pet Azulene Bluing Shampoo—Levlad, 1
My Pet Herbal Shampoo—Levlad, 1
Organimals Organic Pet Shampoo—Aubrey
   Organics, 1
Royal Herbal Pet Shampoo—Pet Connection, 2
Silk Protein Shampoo Concentrate—Levlad, 1

Home Shopping/Mail Order

Best Friend Dog Soap—Brookside Soap, 1
Herbal Pet Shampoo—Ecco Bella, 1
Herbal Pet Shampoo—Naturally Yours, Alex, 1

PETROLEUM JELLY

General Retail Stores

Vegelatum Pure Vegetable Gelly—Mountain
   Fresh Products, 1

Health Food Stores

Baby Un-Petroleum Jelly—Autumn Harp, 1
Un-Petroleum Jelly—Autumn Harp, 1

PLANT CARE PRODUCTS

General Retail Stores _____

Miracid—Stern's Miracle-Gro Products, Inc., 1
Miracle-Gro—Stern's Miracle-Gro Products, Inc., 1
Miracle-Gro for Roses—Stern's Miracle-Gro
  Products, Inc., 1
Miracle-Gro for Tomatoes—Stern's Miracle-Gro
  Products, Inc., 1
Miracle-Gro Lawn Food—Stern's Miracle-Gro
  Products, Inc., 1
Miracle-Gro No Clog Garden & Lawn Feeder—
  Stern's Miracle-Gro Products, Inc., 1
Miracle-Gro Patio Plant Food—Stern's Miracle-Gro
  Products, Inc.,1
Miracle-Gro Therapy—Stern's Miracle-Gro
  Products, Inc., 1

Home Shopping/Mail Order _____

Basic-H Concentrated Soil Conditioner—Shaklee
  U.S., Inc., 1
Sea-Min—Wachter's, 1
Sea-Spraa—Wachter's, 1

PLASTIC AND VINYL CLEANERS

General Retail Stores _____

Plastic Cleaner Magic—Magic American Corp., 1
Vinyl Magic—Magic American Corp., 1

POT AND PAN CLEANERS

General Retail Stores _____

Earth Friendly Pot and Pan Cleaner—Venus
  Laboratories, Inc., 1

POTPOURRI AND SACHETS

General Retail Stores

Enviromental Fragrances Potpourri—TerraNova, 1
Fragrance Puffs—Botanicus, 1
Potpourri—Botanicus, 1

Health Food Stores

Potpourri—No Common Scents, 1

Home Shopping/Mail Order

All potpourri and sachets—Key West Fragrance &
    Cosmetic Factory, Inc., 1
Potporri—Jeanne Rose, 2

Specialty Stores

Acetate Cubes—Scarborough and Company, 1
Enviromental Oils—Scarborough and Company, 1
Potpourri—Potions & Lotions, 3
Potpourri—Scarborough and Company, 1
Sachets—Potions & Lotions, 3
Sachets—Scarborough and Company, 1
Simmering Potpourri—Potions & Lotions, 3
Simmering Spices—Scarborough and Company, 1

POWDER

General Retail Stores

Botanicals Talcum—Smith & Vandiver, Inc., 1
China Lily Body Powder—TerraNova, 1
China Mist Body Powder—TerraNova, 1
Djer-Kiss Talcum—A.I.I. Clubman, 1
Jojoba Body Talc—Botanicus, 1

Mavis Talcum—A.I.I. Clubman, 1
Moon Shadows Body Powder—TerraNova, 1
Pikai Body Powder—TerraNova, 1
Rain Body Powder—TerraNova, 1

Health Food Stores

Body Powder—Dr. Hauschka Cosmetics, 1
Ida Grae Earth Translucent Powder—Nature's
    Colors, 2
Moonsilk Body Powder—Alexandra Avery, 1
Natural Body Powder—Aura Cacia, 1
White Balsam Body Power—Heritage Store, 1
Yolanda Talcum Powder—Beauty Without Cruelty,
    Ltd., 1

Home Shopping/Mail Order

Aloe Body Powder—Grace Cosmetics/Pro-Ma
    Systems, 1
Exquisite Dusting Powder—Mary Kay Cosmetics,
    Inc., 3
L'Arome Echoes Talcum Powder—L'Arome USA,
    Inc., 1
Silk Dust Body Powder—Oriflame International, 3
Talc-Free Body Powder—Ecco Bella, 1

Specialty Stores

Damask Rose Talcum Powder—Crabtree &
    Evelyn, Ltd., 1
Dusting Powder—Body Shop (Calif.), 1
Dusting Powders—Crabtree & Evelyn, Ltd., 1
Lavender Talcum Powder—Crabtree & Evelyn,
    Ltd., 1
Lily of the Valley Talcum Powder—Crabtree &
    Evelyn, Ltd., 1
Millefleurs Talcum Powder—Crabtree & Evelyn,
    Ltd., 1
Perfumed Body Powder—Scarborough and
    Company, 1

Sandalwood Talcum Powder—Crabtree & Evelyn, Ltd., 1
Vetiver Talcum Powder—Crabtree & Evelyn, Ltd., 1

RAZORS AND DEPILATORIES

General Retail Stores

Bonded Razor Set—Wilkinson Sword, 1
Chromium Double Edge Shaving Razors—Wilkinson Sword, 1
Classic Double Edge Shaving Razor—Wilkinson Sword, 1
Colours Disposable Shaving Razors—Wilkinson Sword, 1
Colours Sensitive Skin Shaving Razors—Wilkinson Sword, 1
Profile Shaving Razors—Wilkinson Sword, 1
Shava II Disposable Shaving Razors—Wilkinson Sword, 1
Slim Line Personal Shaver—Ardell International, 2
Surgi-Cream—Ardell International, 2
Surgi-Cream Bikini and Leg Area—Ardell International, 2
Surgi-Cream Brow Shapers Epilating Strips—Ardell International, 2
Surgi-Cream Lift-Off—Ardell International, 2
System II Shaving Razor—Wilkinson Sword, 1

Ultra Glide Disposable Razors—Wilkinson Sword, 1
Ultra Glide Shaving Razors—Wilkinson Sword, 1

Home Shopping/Mail Order

Dry Skin & Hair Remover Mitt—BabyTouch, 1

Specialty Stores

Dry Shaving Stick—Crabtree & Evelyn, 1
Razor Strop—Crabtree & Evelyn, Ltd., 1
Straight Razor—Crabtree & Evelyn, Ltd., 1
Strop Pastel—Crabtree & Evelyn, Ltd., 1

RUST REMOVERS

General Retail Stores

Magic Rust Stain Remover—Magic American
  Corp., 1
Rit Rust Remover—Best Foods, 1

Home Shopping/Mail Order

Like Nu Rust Remover—AFM Enterprises, Inc., 1

SCALP TREATMENTS

General Retail Stores —————————————

Hofels Nettle Hair Tonic—Baudelaire, 1
Jojoba Hair and Scalp Conditioner—A.I.I.
Clubman, 1

Hair Salons —————————————

Active Formulas—Aveda Corp., 1
Balancing Infusion—Aveda Corp., 1
Calming Nutrients—Aveda Corp., 1

Health Food Stores —————————————

Hairever Cleansing Scalp Treatment—Home Health
Products, 2
Hairever Hair & Scalp Vitamin Tonic—Home
Health Products, 2
Neem Hair Lotion—Dr. Hauschka Cosmetics, 1
Neem Hair Oil—Dr. Hauschka Cosmetics, 1
Rosemary Hair Oil—Alexandra Avery, 1

Home Shopping/Mail Order —————————————

Beautifying Hair and Scalp Nourisher—Ecco
Bella, 1
Energizing Hair and Scalp Treatment—G. T.
International, 1

Let-It-Grow Hair and Sculpting Nourisher—Ecco
    Bella, 1
Scalp Tonic—Shirley Price Aromatherapy, 1

SCAR TISSUE PRODUCTS

Health Food Stores

Scarmassage—Heritage Store, 2
Scarmassage Cream—Heritage Store, 2

Home Shopping/Mail Order

Scar Tissue Cream—Shirley Price Aromatherapy, 1

SEALERS AND PRIMERS

Home Shopping/Mail Order

Donnos-Wood Pitch Impregnation—Livos Plant
    Chemistry, 1
Dubno-Primer Oil—Livos Plant Chemistry, 1
Duro-Metal Primer—Livos Plant Chemistry, 1
Dyno Flex—AFM Enterprises, Inc., 1
Dyno Seal—AFM Enterprises, Inc., 1
Hard Seal—AFM Enterprises, Inc., 1
Klear Seal—AFM Enterprises, Inc., 1
Linus-Linseed Impregnation—Livos Plant
    Chemistry, 1
Meldos-Hard Sealer—Livos Plant Chemistry, 1
Menos-Primer Paint—Livos Plant Chemistry, 1
Penetrating Water Seal—AFM Enterprises, Inc., 1
Polyuraseal—AFM Enterprises, Inc., 1

Safecoat Primer Undercoater—AFM Enterprises, Inc., 1
Shingle Protek—AFM Enterprises, Inc., 1
Vinyl Block—AFM Enterprises, Inc., 1
Water Seal—AFM Enterprises, Inc., 1

SELF-TANNING PRODUCTS

General Retail Stores

Solar-Free Body Bronzer—Clientele, 1

Hair Salons

Selftan—Mastey de Paris, 1

Health Food Stores

Perfect Color Self-Tanning Lotion—Jason Natural Products, 1
Sans Sun Face Bronzer—Zia Cosmetics, 1
Sans Sun Tanning Creme— Zia Cosmetics, 1
Sunless Bronze—Borlind of Germany, 1
Tan Without Sun—Reviva Labs, 3

Home Shopping/Mail Order

Key West Bronzer—Key West Fragrance & Cosmetic Factory, Inc., 1
Natural Self-Tanning Complex—Gruene, 1
No Sun—World of Aloe, 1
Self Tanning Creme—La Costa Products International, 1
Self Tanning Facial Moisturizer—La Costa Products International, 1
Sunless Tan Formula—Michel Constantini Natural Cosmetics, 2

Tan Anytime Moisturizing Lotion—Sombra, 1
Translucent Bronzer—Body Shop, 2

Specialty Stores

Sunless Tan—Potions & Lotions, 3

SHAMPOOS

General Retail Stores

Aloe Honey-Rich Shampoo—Nature de France, 3
Alternate Day Shampoo—Botanicus, 1
Apiana Honey Shampoo—Baudelaire, 2
Apiana Propolis Shampoo—Baudelaire, 2
Botanicals Shampoo—Smith & Vandiver, Inc., 1
Chamomile Protein Shampoo—Nature de France, 3
Citre Shine—Advanced Research Labs, 1
Coconut Protein Shampoo—TerraNova, 3
Curlex Shampoo—Dena Corp., 1
Deep Cleansing Shampoo—SafeBrands, Inc., 1
Egyptian Henna Hot Oil Conditioning—A.I.I.
     Clubman, 1
Egyptian Henna Neutral Conditioning Shampoo—
     A.I.I. Clubman, 1
European Mystique Shampoo—Dena Corp., 1
Extra Body Shampoo—SafeBrands, Inc., 1
Extra Gentle Shampoo—SafeBrands, Inc., 1
Faith In Nature Jojoba Shampoo—Baudelaire, 1
Faith In Nature Rosemary Shampoo—Baudelaire, 1
Faith In Nature Seaweed Shampoo—Baudelaire, 1
Faith In Nature Shampoo—Baudelaire, 1
Gee, Your Hair Smells Terrific—Andrew Jergens, 3
*Golden Lotus Jojoba Creme Conditioning
     Shampoo*—Mountain Fresh Products, 1
Golden Lotus Lemongrass Shampoo—Mountain
     Fresh Products, 1

Golden Lotus Lotus Shampoo—Mountain Fresh Products, 1
Golden Lotus Silk Protein Shampoo—Mountain Fresh Products, 1
Hair Masque Shampoo—Clientele, 1
Herbal Shampoo—Abbaco, Inc., 3
Herbal Shampoo—Mountain Fresh Products, 1
Hofels Clove & Dandelion Shampoo—Baudelaire, 1
Hofels Nettle & Rosemary Shampoo—Baudelaire, 1
Hofels Peach & Almond Shampoo—Baudelaire, 1
Its Organic Naturally Shampoo—Dena Corp., 1
Jacob Hooy Arnica Shampoo—Baudelaire, 1
Jacob Hooy Chamomile Shampoo—Baudelaire, 1
Jacob Hooy Lavender Shampoo—Baudelaire, 1
Jacob Hooy Nettle Shampoo—Baudelaire, 1
Jojoba Energizing Shampoo—A.I.I. Clubman, 1
Lectrify Shampoo—Dena Corp., 1
Meta 1 Step—Dena Corp., 1
Moisturizing Shampoo—Botanicus, 1
Natural Shampoo—Tom's of Maine, 1
Normal Shampoo—SafeBrands, Inc., 1
Pikaki Shampoo—TerraNova, 3
Rain Shampoo—TerraNova, 3
Smash Shampoo—Dena Corp., 1
Smith & Vandiver Shampoo—Smith & Vandiver, Inc., 1
Wysong Natural Revitalizing Shampoo—Wysong Corp., 1
Wysong Rinseless Shampoo—Wysong Corp., 1

Hair Salons

Balance Shampoo—Kenra Naturals, 1
Biojoba Shampoo—JOICO Laboratories, 1
Biokosma Shampoo—Dr. Grandel, Inc., 2
Biotane Shampoo—L'anza, 1
Black Malva Shampoo—Aveda Corp., 1
Blue Malva Shampoo—Aveda Corp., 1
Camomile Shampoo—Aveda Corp., 1
Cello Shampoo—Sebastian International, Inc., 1

Clarte Shampoo—Mastey de Paris, 1
Classic Silver—KMS Research, 1
Cleanse-pHree—KMS Research, 1
CliniDan—KMS Research, 1
Clove Shampoo—Aveda Corp., 1
Econome Shampoo—Mastey de Paris, 1
Enove Shampoo—Mastey de Paris, 1
Enviro-Tek Shampoo—Focus 21, 1
Farmavita Colorsafe Shampoo—Chuckles, Inc., 1
Hair Toys Shampoo—Focus 21, 1
Hydraplex Moisturizing Shampoo—Paul Mazzotta,
    Inc., 1
Jojoba Shampoo—Focus 21, 1
Jojoba Shampoo Treatments—Institute of
    Trichology, 3
Kerapro Shampoo—JOICO Laboratories, 1
Lav'ei Shampoo—JOICO Laboratories, 1
Lavenda Shampoo—L'anza, 1
Lime Supreme Shampoo—House of Lowell, Inc., 3
Madder Root Shampoo—Aveda Corp., 1
Moisture Shampoo—Sebastian International, Inc., 1
Moisturizing Shampoo—KMS Research, 1
NEFA Shampoo—KMS Research, 1
Normal to Dry Hair Shampoo—Focus 21, 1
Paul Mitchell Awapuhi Shampoo—John Paul
    Mitchell Systems, 1
Paul Mitchell Shampoo One—John Paul Mitchell
    Systems, 1
Paul Mitchell Shampoo Three—John Paul Mitchell
    Systems, 1
Paul Mitchell Shampoo Two—John Paul Mitchell
    Systems, 1
Poly-Pro Shampoo—House of Lowell, Inc., 3
Professional Supreme Shampoo—House of Lowell,
    Inc., 3
Release Shampoo—Paul Mazzotta, Inc., 1
Remede Shampoo—L'anza, 1
Resolve Shampoo—JOICO Laboratories, 1
SeaPlasma Shampoo—Focus 21, 1

Shampoo for Dry Hair—Zinzare International,
  Ltd., 2
Shampoo for Normal Hair—Zinzare International,
  Ltd., 1
Sukesha All-Nutrient Shampoo—Chuckles, Inc., 1
Sukesha Extra Body Shampoo—Chuckles, Inc., 1
Sukesha Moisturizing Shampoo—Chuckles, Inc., 1
Sukesha Natural Balance Shampoo—Chuckles,
  Inc., 1
Super Naturals Aloe Vera Shampoo—North
  Country Naturals, 2
Super Naturals Jojoba Shampoo—North Country
  Naturals, 2
Super Naturals Panthenol Shampoo—North
  Country Naturals, 2
Super Naturals Vitamin E Shampoo—North
  Country Naturals, 2
Traite—Mastey de Paris, 1
UltraVolume—KMS Research, 1
Vitro Shampoo—L'anza, 1
Wild Cherry Supreme Shampoo—House of Lowell,
  Inc., 3

Health Food Stores

Aloe Vera Conditioning Shampoo—Orjene Natural
  Cosmetics, 1
Aloe Vera Shampoo—Jason Natural Products, 2
Aloegen Biogenic Treatment Shampoo—Levlad, 3
Aloegen Biotreatment 22 Shampoo—Levlad, 3
Aloegen 60/80 Treatment Shampoo—Levlad, 3
Aloe-Herb Shampoo—Natural Bodycare, 1
Arizona Naturals Normal to Dry Shampoo—Arizona
  Natural Resources, Inc., 2
Arizona Naturals Normal/Oily Shampoo—Arizona
  Natural Resources, Inc., 2
Arizona Naturals Tearless Shampoo—Arizona
  Natural Resources, Inc., 2
Aurora Henna Shampoo—Aurora Henna, 1

Biotin Shampoo w/Na P⁻A—Jason Natural Products, 2

Biscayne Seaweed Shampoo—Natural Solutions, 1

Black Walnut Shampoo—Nirvana, 1

Blackmores Almond & Milk Shampoo—Solgar Vitamin Co., 1

Blackmores Marshmallow Shampoo—Solgar Vitamin Co., 1

Blackmores Wild Nettle Shampoo—Solgar Vitamin Co., 1

Blue Camomile Shampoo—Aubrey Organics, 1

Blue Malva Shampoo—Nirvana, 1

Brewer's Yeast Shampoo—Reviva Labs, 2

CamoCare Shampoo—Abkit, Inc., 1

Camomile Luxurious Herbal Shampoo—Aubrey Organics, 1

Chamomile Shampoo—Nirvana, 1

Chamovera Shampoo—Home Health Products, 2

Clove Shampoo—Nirvana, 1

Conditioning Shampoo—Beehive Botanicals, 2

Cool Alaska Herbal Shampoo—Natural Solutions, 1

Crudoleum Shampoo—Heritage Store, 1

Dr. Gomez Jojoba Shampoo—Desert Essence, 1

E-Gem Organic Shampoo—Carlson Laboratories, 3

E.F.A. Primrose Shampoo—Jason Natural Products, 2

Egyptian Henna Shampoo—Aubrey Organics, 1

Eva Jon Shampoo—Eva Jon Cosmetics, 3

Ginkgo Shampoo—Aubrey Organics, 1

Ginkgo Shampoo—Jason Natural Products, 2

Ginseng Shampoo—Aubrey Organics, 1

Hair Maximum Shampoo—Mountain Ocean, 1

Hair Treatment Shampoo—Earth Science, 1

Hawaiian Seaweed Shampoo—Reviva Labs, 2

Henna Highlights Shampoo—Jason Natural Products, 2

Herbal Astringent Shampoo—Earth Science, 1

Herbal Shampoo—Jason Natural Products, 2

Hibiscus Shampoo—Nirvana, 1

Peppermint Hops Shampoo—Paul Penders, 1
Polynatural 60/80 Shampoo—Aubrey Organics, 1
Primrose & Lavender Herbal Shampoo—Aubrey
    Organics, 1
Pure Herbal Shampoo—Orjene Natural
    Cosmetics, 1
QBHL Quillaya Bark Hair Lather—Aubrey
    Organics, 1
Quite Natural Daily Shampoo—Natural Bodycare, 1
Rainbow Henna Shampoo—Rainbow Research, 1
Rosa Mosqueta Herbal Shampoo—Aubrey
    Organics, 1
Rosemary Lavender Shampoo—Paul Penders, 1
Salon Naturals Highlighting Shampoo—ShiKai
    Products, 1
Santa Ana Lemon Shampoo—Natural Solutions, 3
Saponin A.A.C. Shampoo—Aubrey Organics, 1
Sea Kelp Shampoo—Jason Natural Products, 2
Sea Minerals Shampoo—Sea Minerals, 1
Seide Shampoo—Borlind of Germany, 1
Selenium Natural Blue Shampoo—Aubrey
    Organics, 2
ShiKai Shampoo—ShiKai Products, 3
60/80 Treatment Shampoo—Levlad, 3
Sonora Desert Jojoba Shampoo—Natural
    Solutions, 3
Specialized Black Hair Care—Eva Jon Cosmetics, 3
Sports Shampoo—Rainbow Research Corp., 1
Stoneybrook Botanicals Oil Free Shampoo—
    Rainbow Research Corp, 1
Ultra Hair Shampoo—Eva Jon Cosmetics, 3
Vitamin E Shampoo—Jason Natural Products, 2
Walnut Oil Shampoo—Paul Penders, 1

Home Shopping/Mail Order

Active-Aloe Hair & Scalp Cleanser—Kallima
    International, Inc., 1
Almond Cream Shampoo—Sombra, 1

Aloe Shampoo—Grace Cosmetics/Pro-Ma Systems, 1

Aromatherapy Balancing Shampoo—Ecco Bella, 1

Balancing Shampoo—Mary Kay Cosmetics, Inc., 3

Botanical Shampoo—Wachter's, 1

Chamomile Shampoo—Body Shop, 2

Chamomile Shampoo—Jeanne Rose, 2

Chamomile Shampoo—Michel Constantini Natural Cosmetics, 2

Chamomile Shampoo—O'Naturel, 1

Cherry Almond Shampoo—Sombra, 1

Coconut Oil Shampoo—Body Shop, 2

Conditioning Shampoo—Bronson Pharmaceuticals, 3

Dry & Damaged Shampoo—Wachter's, 1

Dry Hair Shampoo—Jeanne Rose, 2

Finelle Shampoo—Finelle Cosmetics, 2

Frequent Wash Grapefruit Shampoo—Body Shop, 2

Hair Lover's Energizer Shampoo—Hobe Laboratories, Inc., 1

Henna & Rosemary Shampoo—O'Naturel, 1

Henna Cream Shampoo—Body Shop, 2

Herbal Shampoo—Jacklyn Cares, 1

Herbal Shampoo—Jeanne Rose, 2

Highlight Shampoo—Michel Constantini Natural Cosmetics, 2

Honey & Sage Shampoo—O'Naturel, 2

Ice Blue Shampoo—Body Shop, 2

In Depth Shampoo—Key West Fragrance & Cosmetic Factory, Inc., 1

Jojoba Oil Conditioning Shampoo—Body Shop, 2

Jojoba Shampoo—KSA Jojoba, 1

Kelp Protein Shampoo—Jacklyn Cares, 1

Key West Aloe Shampoo—Key West Fragrance & Cosmetic Factory, Inc., 1

L'Arome Echoes Shampoo—L'Arome USA, Inc., 1

Lavender Conditioning Shampoo—O'Naturel, 1

Lusteriz—Magic of Aloe, 3

Maxicore Volumizer Shampoo—La Costa Products International, 1

Walnut Leaf Shampoo—Michel Constantini Natural
Cosmetics, 2

Specialty Stores

Almond Shampoo—Crabtree & Evelyn, Ltd., 1
Aloe Vera Shampoo—Crabtree & Evelyn, Ltd., 1
Apricot Shampoo—Crabtree & Evelyn, Ltd., 1
Avocado Shampoo—Crabtree & Evelyn, Ltd., 1
Birch Shampoo—Crabtree & Evelyn, Ltd., 1
Buttermilk Shampoo—Crabtree & Evelyn, Ltd., 2
Camomile Shampoo—Crabtree & Evelyn, Ltd., 1
Camomile Shampoo—Body Shop (Calif.), 3
China Rain Shampoo—Body Shop (Calif.), 3
Coconut Shampoo—Crabtree & Evelyn, Ltd., 1
Conditioning Shampoo—Body Shop (Calif.), 1
Egg Shampoo—Crabtree & Evelyn, Ltd., 2
Goatmilk Shampoo—Crabtree & Evelyn, Ltd., 2
Henna Shampoo—Crabtree & Evelyn, Ltd., 1
Jojoba Oil Shampoo—Body Shop (Calif.), 3
Jojoba Shampoo—Crabtree & Evelyn, Ltd., 1
Lemon Shampoo—Crabtree & Evelyn, Ltd., 1
Milk & Honey Shampoo—Potions & Lotions, 3
Millefleurs Shampoo—Crabtree & Evelyn, Ltd., 1
Moisturizing Shampoo—Body Shop (Calif.), 3
Nantucket Briar Shampoo—Scarborough and
Company, 2
Protein Shampoo—Body Shop (Calif.), 3
Pure Shampoo—Potions & Lotions, 3
Rosemary Shampoo—Crabtree & Evelyn, Ltd., 1
Sandalwood Shampoo—Crabtree & Evelyn, Ltd., 1
Savannah Garden Shampoo—Scarborough and
Company, 1
Spring Rain Shampoo—Scarborough and
Company, 1
Tiare Tahiti—Crabtree & Evelyn, Ltd., 1
Wheatgerm Shampoo—Crabtree & Evelyn, Ltd., 1
White Ginger Shampoo—Crabtree & Evelyn, Ltd., 1

SHAVING CREAM AND SKIN PRODUCTS

General Retail Stores

After Shave Balm—A.I.I. Clubman, 1

Aloe Vera Shaving Gel—TerraNova, 1

Botanical Shaving Lotion—Smith & Vandiver, Inc., 1

Clubman Shave Cream—A.I.I. Clubman, 1

Craig Martin Mentholated Brushless—Comfort Manufacturing Co., 1

Surrey Shaving Soap—Surrey, Inc., 3

Tom's of Maine Natural Shaving Cream—Tom's of Maine, 1

Wysong Shaving Gel—Wysong Corp., 1

Health Food Stores

After Shave Balm—Alba Botanica, 1

Almond Treatment Scrub Pre-Shave—Earth Science, 1

Aloe/Herbal After Shave Skin Soother—Earth Science, 1

Azulene Moisturizing Shave Creme—Earth Science, 1

Cool Mint Moisture Shave—Kiss My Face, 1

Daily Workout Shave Cream—Natural Solutions, 3

Daily Workout Shave Gel—Natural Solutions, 3

Dry Skin Intensive Treatment Beard Softener—Earth Science, 1

Herbal Mint & Ginseng Shaving Cream—Aubrey Organics, 1

Jasmine Moisture Shave—Kiss My Face, 1

Key Lime Moisture Shave—Kiss My Face, 1

Lemon Balm Shave Cream—Paul Penders, 1

Moisture After Shave Lotion—Paul Penders, 1

Natural After Shave Skin Soother—Orjene Natural Cosmetics, 2

Natural Shaving Cream—Orjene Natural Cosmetics, 2

Organic Formula Cream Shave—Alba Botanica, 2
Shaving Cream—Jason Natural Products, 2
Soothing Shave Cream—Zia Cosmetics, 1

Home Shopping/Mail Order

Aftershave Balm—Shirley Price Aromatherapy, 1
Aloe Cream Shave—Gruene, 1
Aloe Shaving Lotion—Grace Cosmetics/Pro-Ma
Systems, 1
Mr. Aloe Shaving Gel— Magic of Aloe, 3
Shave Cream—Mary Kay Cosmetics, Inc., 3
Shaving Cream—Body Shop, 2

Specialty Stores

After Shave Balm—Crabtree & Evelyn, Ltd., 1
Almond Shaving Cream—Crabtree & Evelyn,
Ltd., 1
Lavender Shaving Cream—Crabtree & Evelyn,
Ltd., 1
Men's Range Shaving Cream—Crabtree & Evelyn,
Ltd., 1
Men's Range Shaving Soap—Crabtree & Evelyn,
Ltd., 1
Sandalwood Shaving Cream—Crabtree & Evelyn,
Ltd., 1
Sandalwood Shaving Soap—Crabtree & Evelyn,
Ltd., 1
Shave Gel with Aloe Vera—Scarborough and
Company, 3
Shave Gel—Scarborough and Company, 1
Shaving Gel—Body Shop (Calif.), 2

SHOE POLISH

General Retail Stores

Camp Dry Aerosol—Kiwi Brands, Inc., 3

Cavalier Aerosols—Kiwi Brands, Inc., 3
Cavalier Desalter—Kiwi Brands, Inc., 3
Cavalier Ever-Dri Liquid—Kiwi Brands, Inc., 3
Cavalier Liquid and Cream Polishes—Kiwi Brands, Inc., 3
Esquire Cleaner Conditioner—Kiwi Brands, Inc., 3
Esquire Paste and Liquid Polishes—Kiwi Brands, Inc., 3
Esquire Patent & Pattina Cleaner—Kiwi Brands, Inc., 3
Kiwi Aerosols—Kiwi Brands, Inc., 3
Kiwi Golf Shoe Care Products—Kiwi Brands, Inc., 3
Kiwi Honor Guard Products—Kiwi Brands, Inc., 3
Kiwi Parade Gloss—Kiwi Brands, Inc., 3
Kiwi Paste, Liquid, and Creme Polishes—Kiwi Brands, Inc., 3
Kiwi Saddle Soap—Kiwi Brands, Inc., 3
Kiwi Shoe Patch—Kiwi Brands, Inc., 3
Kiwi Sneaker Care Products—Kiwi Brands, Inc., 3
Kiwi Sneaker Fresh—Kiwi Brands, Inc., 3
Kiwi Sport Products—Kiwi Brands, Inc., 3
Kiwi Water Repellent Pastes and Liquids—Kiwi Brands, Inc., 3
Tana Aerosols—Kiwi Brands, Inc., 3
Tana Liquid, Cream, and Paste Polishes—Kiwi Brands, Inc., 3
Tana Style Products—Kiwi Brands, Inc., 3

Home Shopping/Mail Order _____

Bertos-Leather Seal—Livos Plant Chemistry, 1
E-Z On Shoe Polish—AFM Enterprises, Inc., 1
Snado-Leather & Shoe Polish—Livos Plant Chemistry, 1

SKIN AND BODY OILS

General Retail Stores _____

Aromatic Skin Oils—Gregory, 1

Essential Body Oils—Faith in Nature, 1
Face & Body Natural Oil Moisturizer—La Crista,
 Inc., 1
Fragylis Precious Oils—Jean Pax, Inc., 1
100% Pure Coconut Oil—Mountain Fresh
 Products, 1

Hair Salons

Active Formula—Aveda Corp., 1
Balancing Infusions—Aveda Corp., 1
Calming Nutrients—Aveda Corp., 1
Energizing Nutrients—Aveda Corp., 1
Love—Aveda Corp., 1

Health Food Stores

A and E Oil—Michael's Health Products, 1
Almond Oil—Heritage Store, 1
Aloe Vera Beauty Oil—Jason Natural Products, 1
Apricot Body and Facial Oil—Earth Science, 1
Apricot Kernel Moisturizing Oil—Orjene Natural
 Cosmetics, 1
Aussie Gold 100% Pure Tea Tree Oil—Jason
 Natural Products, 1
Avocado 7 Oils—Orjene Natural Cosmetics, 3
Blackmores Apricot E Oil—Solgar Vitamin Co., 1
Body Oil Blackthorn Composition—Dr. Hauschka
 Cosmetics, 1
Body Oils—Aroma Vera, 1
Citrus Body Oil—Weleda, Inc., 1
E-Gems Oil Drops—Carlson Laboratories, 1
Facial Oils—Aroma Vera, 1
Facial Skin Oil—Dr. Hauschka Cosmetics, 1
Garlic and Goldenseal Oil—Michael's Health
 Products, 1
Glowing Touch Almond Skin Care Oils—Sunshine
 Products Group, 1
Hawaiian Kukui Nut Oil—Orjene Natural
 Cosmetics, 1

Jungle Blossom Body Oil—Alexandra Avery, 1
Mothers Special Blend Skin Toning Oil—Mountain
    Ocean, 1
Mountain Herb Body Oil—Alexandra Avery, 1
Skin Savior Oil—WiseWays Herbals, 1
Sweet Almond Oil—Orjene Natural Cosmetics, 1
White Pond Lily and Wild Rose Skin Care Oil—
    Lakon Herbals, 1

Home Shopping/Mail Order

Apricot Kernel Oil—Body Shop, 2
Body Supple Dry Oil Spray—Oriflame
    International, 3
Carrot Facial Oil—Body Shop, 2
Essential ReNutrient Skin Oils—Kallima
    International, Inc., 1
Facial Fortifier Oil—Oriflame International, 3
Facial Oils—Simplers Botanical Co., 1
Natural Golden Oils—Jacklyn Cares, 1
Sweet Almond Oil—Body Shop, 2
Wheatgerm Oil—Body Shop, 2

Specialty Stores

Natural Skin Oil—Body Shop (Calif.), 1
Skin Radiance Oil Spray—Scarborough and
    Company, 1
Super Oil—i Natural Skin Care and Cosmetics, 1

SKIN BLEACHING PRODUCTS

General Retail Stores

Invisi-Bleach—Ardell International, 2
Jolen Creme Bleach—Jolen, Inc., 1

Health Food Stores ───────────────────

Brown Spot Cream—Reviva Labs, 2

Home Shopping/Mail Order ─────────────────

Banishing Cream—Avon Products, Inc., 3

SKIN FRESHENERS

General Retail Stores ───────────────────

Aloe Vera Gentle Freshener—TerraNova, 1
Dry Skin Freshener—Colonial Dames Co., Ltd., 1
Pore Minimizing Skin Freshener—Cosmyl
  Cosmetics, 3
Skin Freshener—Colonial Dames Co., Ltd., 1
Vitamin E Toning Freshener—Colonial Dames Co.,
  Ltd., 1

Hair Salons ───────────────────

Pure Rosewater Freshener—Jurlique D'Namis
  Ltd., 1
Super Naturals Elastin Freshener—North Country
  Naturals, 3

Health Food Stores ───────────────────

Aloe Vera Energizer Refreshener—Orjene
  Cosmetics, 1
Aloe-Herb Tonic Freshener/Toner—Natural
  Bodycare, 1
Apricot Kernel Oil Freshener—Orjene Natural
  Cosmetics, 1
Cool Peppermint Freshener—Beauty Without
  Cruelty, 1

Gentle Camomile Freshener—Beauty Without
  Cruelty, 1
Rose Petal Skin Freshener—Beauty Without
  Cruelty, 1

Home Shopping/Mail Order

Alcohol-Free Refreshener—Shaklee U.S.,
  Inc., 1
Aloe pH Freshener—Grace Cosmetics/Pro-Ma
  Systems, 1
Desert Mist Freshener—Hobe Laboratories, Inc., 1
French Rosewater pH Balancing Skin Freshener—
  Indian Creek, 1
Freshener—World of Aloe, 1
Gentle Action Freshener—Mary Kay Cosmetics,
  Inc., 3
Grapefruit Facial Freshener—Jacklyn Cares, 1
Grapefruit Freshener—Barbizon International, 3
Herbal Refreshener—Shaklee U.S., Inc., 1
Lemon Skin Freshener—Sombra, 1
Moisan Key Tone Freshener—Nutri-Metrics
  International, 1
Refining Freshener—Mary Kay Cosmetics, Inc., 3
Refresh—Ecco Bella, 1
Skin Tone Freshener—Magic of Aloe, 3
Wild Fern Freshener—Patricia Allison, 1

Specialty Stores

Aloe Vera Freshener—Body Shop (Calif.), 1
Grapefruit Freshener—i Natural Skin Care and
  Cosmetics, 1
Refresher—Body Shop (Calif.), 1
Rosewater Freshener—Body Shop (Calif.), 1

SKIN PROTECTORS

Health Food Stores

Warm Skin Winterguard—Aurora Henna, 2

Winterizer Skin Protection Balm—Aubrey
   Organics, 1

SPECIALITY CLEANERS

General Retail Stores

Bar Keepers Friend Cleaner—SerVaas
   Laboratories, 1
Chandelier Magic—Magic American Corp., 1
Fiberglass Magic—Magic American Corp., 1
Garage Magic—Magic American Corp., 1
Goo Gone—Magic American Corp., 1
Marble Magic—Magic American Corp., 1
Microwave Oven Magic—Magic American Corp., 1

Specialty Stores

Vacuum Refresher—Scarborough and Company, 1

STAINS, SHELLACS, AND VARNISHES

Home Shopping/Mail Order

Acrylacq—AFM Enterprises, Inc., 1
Bela Wood Stain—Livos Plant Chemistry, 1
Duro Stain—AFM Enterprises, Inc., 1
Earthen & Mineral Stain Pastes—Livos Plant
   Chemistry, 1
Kaldet Resin & Oil Finish—Livos Plant
   Chemistry, 1
Landis-Shellac—Livos Plant Chemistry, 1

Safecoat Water Based Wood Stain—AFM
   Enterprises, Inc., 1
Taya-Wood Glaze—Livos Plant Chemistry, 1
Trebo-Shellac—Livos Plant Chemistry, 1
Tunna Furniture Varnish—Livos Plant
   Chemistry, 1

STARCH AND SIZING

General Retail Stores

Argo Gloss Laundry Starch—Best Foods, 1
Niagara Sizing Fabric Finish—Best Foods, 1
Niagara Spray Starch—Best Foods, 1

SUNBURN CARE PRODUCTS

Health Food Stores

Soothing Herbal After-Sun Spray—Earth Science, 1

Home Shopping/Mail Order

Great Stuff Sunburn Relief—Hobe Laboratories,
   Inc., 1
Sun*Burn Solution—Simplers Botanical Co., Inc.

SUNTAN ACCELERATORS

Hair Salons

Bronzee Tanning Accelerator—Focus 21, 1

Paul Mitchell The Accelerator—John Paul Mitchell Systems, 1

Health Food Stores

PreTan Activator—Reviva Labs, 2
Sunlind Tropic—Borlind of Germany, 2
Tan Up Natural Accelerator—Aubrey Organics, 1

Specialty Stores

Natural Protection Suntan Accelerator—i Natural Skin Care and Cosmetics, 3

SUNTAN LOTIONS AND BLOCKS

General Retail Stores

Anti-Aging Sun Stick—Clientele, 1
Children's Sun Block—Wysong Corp., 1
Ion Protector—Clientele, 1
Premiere Sun Block—Tyra Skin Care, 2
Solar Protector Lotion—Clientele, 1
Wysong Sun Block—Wysong Corp., 1
Wysong Sun Screen—Wysong Corp., 1
Wysong Tanning Lotion—Wysong Corp., 1

Hair Salons

Activan—Mastey de Paris, 1
Biokosma Sun Cream—Dr. Grandel, Inc., 1
Biokosma Sun Protection Foam—Dr. Grandel, Inc., 1
Biokosma Tibetan Sun Care—Dr. Grandel, Inc., 1
Bronzee Deep Oil Spray—Focus 21, 1
Bronzee Oil Spray—Focus 21, 1
Sun Cream—Jurlique D'Namis Ltd., 1
Sun Emulsion Collagen Protection—Dr. Babor Natural Cosmetics, 3

Sunbloc—Mastey de Paris, 1
Sunblock/Moisturizer—Studio Magic, 2
Suntan Creme—Mastey de Paris, 1

Health Food Stores

Aloe Gardenia Sun Oil—Alexandra Avery, 1
Aloe 'n E Tanning Lotion—Earth Science, 1
Aloe Suma Advanced Treatment—Rachel Perry, 1
Aloe Suma Sunblock—Rachel Perry, 1
Aloe Suma Tanning Formula—Rachel Perry, 1
Arizona Naturals Maximum Protection—Arizona
    Natural Resources, Inc., 2
Arizona Naturals Minimum Protection—Arizona
    Natural Resources, Inc., 2
Arizona Naturals Total Block Waterproof—Arizona
    Natural Resources, Inc., 2
Beach Baby—Reviva Labs, 2
Bio-Pure Suntan Lotion w/Tyrosine—Reviva Labs, 2
Body Butter—Aura Cacia, 1
Duck Oil Waterproof Sunscreen—Jason Natural
    Products, 1
Gentle Tanning Lotion—Orjene Natural
    Cosmetics, 1
Natural Bronze Tanning Oil—Earth Science, 1
Nature Tan—Aubrey Organics, 3
Saving Face—Aubrey Organics, 1
Sun Butter—Aura Cacia, 1
Sun Creams—Borlind of Germany, 2
Sun Gel—Borlind of Germany, 1
Sun Milk—Borlind of Germany, 2
Sunblock—Mountain Ocean, 1
Sunbrellas—Jason Natural Products, 1
Sunbrellas Nose/Lip Guard—Jason Natural
    Products, 1
Sunscreen—Mountain Ocean, 1
Sunshade—Aubrey Organics, 3
Suntan Lotion—Reviva Labs, 2
Super Sun Protector Lotion—Orjene Natural
    Cosmetics, 2

Tropical Balm—Aurora Henna, 1
Ultra 15 Herbal Sunblock—Aubrey Organics, 1

Home Shopping/Mail Order

Aloe Block—Magic of Aloe, 3
Aloe Screen Out Lotion—Grace Cosmetics/Pro-Ma
   Systems, 1
Aloe Tanning Lotion—Grace Cosmetics/Pro-Ma
   Systems, 1
Body Shop Sunscreens—Body Shop, 2
Cocoa Butter Suntan Lotion—Body Shop, 2
Dura Tan—Magic of Aloe, 3
Facial Sunblock—Mary Kay Cosmetics, Inc., 3
4 D Hobe Suntan Lotion—Hobe Laboratories, Inc., 1
Grab A Tan—Key West Fragrance & Cosmetic
   Factory, Inc., 1
Key West Sunkids—Key West Fragrance &
   Cosmetic Factory, Inc., 1
Key West Tan—Key West Fragrance & Cosmetic
   Factory, Inc., 1
Lip Protector—Mary Kay Cosmetics, Inc., 3
Nutrient Balm Sunscreen—Patricia Allison, 1
Oil Free Facial Sunblock—Mary Kay Cosmetics,
   Inc., 3
Oil Free Sun Spray—La Costa Products
   International, 1
Scuba Waterproof Gel—Key West Fragrance &
   Cosmetic Factory, Inc., 1
Sensitive Skin Waterproof Sunblock—Mary Kay
   Cosmetics, Inc., 3
Shaklee Naturals Sunblock—Shaklee U.S., Inc., 1
Society Tan—Magic of Aloe, 3
SPF Cremes—La Costa Products International, 1
Stingray Waterproof Sun Protection—Key West
   Fragrance & Cosmetic Factory, Inc., 1
Sun Care System—Finelle Cosmetics, 2
Sun Sensitive Cream—Irma Shorell, 2
Sun Stick SPF 12—Finelle Cosmetics, 2
Sun-Dew Sun Filter Moisturizer—Wachter's, 1

SunBlock—Michel Constantini Natural
    Cosmetics, 2
Sunsorb—Key West Fragrance & Cosmetic Factory,
    Inc., 1
Super Sunblock—Mary Kay Cosmetics, Inc., 3
Tan Tan Coconut Oil—Key West Fragrance &
    Cosmetic Factory, Inc., 1
Triple Tan—Key West Fragrance & Cosmetic
    Factory, Inc., 1
Waterproof Sunscreen—Mary Kay Cosmetics, Inc., 3
Year Round Sun Shield—Gruene, 1

Specialty Stores

Aloe Vera Tanning Lotion—Body Shop (Calif.), 3
*Natural Protection PABA Free Protective Face
    Cream*—i Natural Skin Care and Cosmetics, 3
Natural Protection Sunscreen—i Natural Skin Care
    and Cosmetics, 3
Sunscreen Lotions—Body Shop (Calif.), 1

SUNTAN MAINTAINANCE PRODUCTS

General Retail Stores

Aqua Protector—Clientele, 1
Solar-Free Face Gel—Clientele, 1

Hair Salons

Biokosma After Sun—Dr. Grandel, Inc., 1
Paul Mitchell After Sun-Tan Extender—John Paul
    Mitchell Systems, 1
Sun Calm—Mastey de Paris, 1

Health Food Stores

After Sun Tan Maintenance & Moisturizer—
Aubrey Organics
*After-sun—*Michael's Health Products, 1
*Sunlind After Sun—*Borlind of Germany, 1

Home Shopping/Mail Order

*Active Tan Enhancing Lotion—*Michel Constantini
Natural Cosmetics, 2
*After Sun Gel—*Finelle Cosmetics, 2
*Save A Tan—*Key West Fragrance & Cosmetic
Factory, Inc., 1
*Sun Treatment—*La Costa Products International, 1

Specialty Stores

*After Sun Milk—*Crabtree & Evelyn, Ltd., 1

Tanning Salons

*NuSun—*NuSun/Marchemco, 1

SWIMMERS' SHAMPOOS

Health Food Stores

*Swimmers & Sport Shampoo—*Jason
Natural Products, 1
*Swimmers Shampoo—*Aubrey Organics, 1

Home Shopping/Mail Order

*Dechlorinating Shampoo—*Michel Constantini
Natural Cosmetics, 2

Specialty Stores

*Swimmer's Shampoo—*Body Shop (Calif.), 3

THINNERS

Home Shopping/Mail Order _____

Kiros-Alcohol Thinner—Livos Plant Chemistry, 1
Leinos-Citrus Thinner—Livos Plant Chemistry, 1

THROAT AND BREAST CREAMS

Hair Salons _____

Silhouette Cream Bust—Dr. Babor Natural
  Cosmetics, 1

Health Food Stores _____

LL Decolleté Cream—Borlind of Germany, 2

Home Shopping/Mail Order _____

Throat & Bust Firmer—Oriflame International, 3
TLC Throat Cream—The Face Food Shoppe, 2
Translucent Firming Throat/Breast—La Costa
  Products, 1
Veeline—Key West Fragrance & Cosmetic Factory,
  Inc., 1

TINTS AND DYES—FABRIC

General Retail Stores

Rit All Purpose Tints and Dyes—Best Foods, 1
Rit Design-It Dye Binder—Best Foods, 1
Rit Fast Fade for Jeans—Best Foods, 1
Tintex Fabric Dyes— Kiwi Brands Inc., 3

TOILET BOWL CLEANERS

General Retail Stores

BLOO Toss-Ins—Kiwi Brands, Inc., 3
Earth Friendly Unibol Blocks—Venus Laboratories, Inc., 1
Earth Friendly Unitab Urinal Blocks—Venus Laboratories, Inc., 1
Earth Rite Toilet Bowl Cleaner—Magic American Corp., 1
Johnny Blue—Kiwi Brands, Inc., 3
Johnny Green—Kiwi Brands, Inc., 3
Toilet Bowl Cleaner—Magic American Corp., 1
TY-D-BOL Power Tabs—Kiwi Brands, Inc., 3

Health Food Stores

Toilet Cleaner—Ecover, 1

Home Shopping/Mail Order

Toilet Cleaner—Seventh Generation, 1

TOOTHPASTE

Dentists

Oxyfresh Toothpaste—Oxyfresh, 1

General Retail Stores

Craig Martin Fluoride Toothpaste—Comfort
Manufacturing Co., 1
Craig Martin GEL Flouride Toothpaste—Comfort
Maufacturing, 1
Craig Martin Milk of Magnesia Toothpaste—Comfort Mfg. Co., 1
Craig Martin MINT Flouride Toothpaste—Comfort
Manufacturing Co., 1
Fluoridated Toothpaste—Adwe Laboratories, 1
Natural Baking Soda Toothpaste—Tom's of
Maine, 1
Natural Toothpaste w/Fluoride—Tom's of Maine, 1
Natural Toothpaste w/Propolis and Myrrh—
Tom's of Maine, 1
Sarakan Toothpaste—Baudelaire, 1

Health Food Stores

Auromere Herbal Toothpaste—Auromere, Inc., 1
Dental Care with Aloe—Eva Jon Cosmetics, 3
Herbal Vedic Toothpaste—Auroma International, 1
IPSAB Tooth Powder—Heritage Store, 1
Mer-flu-an Natural European Toothpowder—
American Merfluan, Inc., 1
Natural Salt Toothpaste—Weleda, Inc., 1
Nature's Gate Natural Gel Toothpaste—Levlad, 1
Nature's Gate Natural Toothpaste—Levlad, 1
Peelu Tooth Powder—Peelu Products, 1
Peelu Toothpaste—Peelu Products, 1
Pink Toothpaste With Myrrh—Weleda, Inc., 1
Plant Gel Toothpaste—Weleda, Inc., 1
Propolis Toothpaste—Beehive Botanicals, 2
Rainbow Natural Toothpaste—Rainbow Research
Corp., 2
Salt 'N Soda—Home Health Products, 2
Tea Tree Oil Toothpaste—Desert Essence, 1
Vicco Herbal Toothpaste—Auromere, Inc., 1

Home Shopping/Mail Order ———————————

Dent-A-Kleen Tooth Gel—Hobe Laboratories, Inc., 1
New Concept Organic Dentifrice—Shaklee U.S.,
Inc., 2

Specialty Stores ———————————

Camomile Toothpaste—Crabtree & Evelyn, Ltd., 1
Herbal Toothpaste—Crabtree & Evelyn, Ltd., 1

TOYS

General Retail Stores ———————————

Hasbro Toys—Hasbro, Inc., 1
Milton Bradley—Hasbro, Inc., 1
Playskool Toys—Hasbro, Inc., 1

Home Shopping/Mail Order ———————————

All toys—Livos Plant Chemistry, 1

Specialty Stores ———————————

All toys—Potions & Lotions, 3

VITAMIN E PRODUCTS

General Retail Stores ———————————

Vitamin E Oil—TerraNova, 1

Hair Salons ————————————————————

Super Naturals Vitamin E Ointment—North
    Country Naturals, 2
Super Naturals Vitamin E Skin Oil—North Country
    Naturals, 1

Health Food Stores ————————————————

Natural Vitamin E Complexion and Body Oil—
    Earth Science, 1
Vitamin E Cream—Jason Natural Products, 1
Vitamin E Cream—Orjene Natural Cosmetics, 3
Vitamin E Enriched Facial Oil—Natural
    Solutions, 1
Vitamin E Lotion—Orjene Natural Cosmetics, 3
Vitamin E Oil—Home Health Products, 2
Vitamin E Oil—Orjene Natural Cosmetics, 1
Vitamin E Oils—Jason Natural Products, 1
Vitamin E—Aroma Vera, 1

Home Shopping Mail Order ————————————

Pure Vitamin E Oil—Bronson Pharmaceuticals, 1
Vitamin E Beauty Cream—Bronson
    Pharmaceuticals, 3
Vitamin E Beauty Oil—Bronson Pharmaceuticals, 3

VITAMINS, MINERALS, AND SUPPLEMENTS

General Retail Stores ————————————————

Daily Nutrients—Clientele, 1
Stress Control Nutrients—Clientele, 1
Wysong Vitamins and Minerals—Wysong
    Medical Corp., 1

Hair Salons

Paul Mazzotta Vegetable Protein Powder—Paul Mazzotta, Inc., 1

Health Food Stores

All vitamins, minerals, and/or supplements—Carlson Laboratories, 3

All vitamins, minerals, and/or supplements—Heritage Store, 1

All vitamins, minerals, and/or supplements—Michael's Health Products, 3

Earth Science Supplements—Earth Science, 1

Herbal Vedic Ayurvedic Formulas—Auroma International, 1

Integrated Health Supplements—Integrated Health, 1

Micellized Liquid Vitamins—Earth Science, 1

New Chapter Food Supplements—New Moon Extracts, Inc., 1

Skin Capsules—Dr. Hauschka Cosmetics, 3

Solgar—Solgar Vitamin Co., 3

Vegetarian Supplements—Royal Laboratories, 1

Home Shopping/Mail Order

All vitamins, minerals, and/or supplements—Biogime, 1

All vitamins, minerals, and/or supplements—Bronson Pharmaceuticals, 3

All vitamins, minerals, and/or supplements—Hobe Laboratories, Inc., 1

Nuhairtrition Vitamins and Minerals—Hobe Laboratories, Inc., 1

Shaklee—Shaklee U.S., Inc., 1

Swedish Supreme Vitamin & Mineral Supplement—Cernitin America, Inc., 1

Vitamins and Nutritional Supplements—Golden Pride/Rawleigh, 2

W

WATER SOFTENERS

Health Food Stores _____

Sal Suds Powder Water Softener—Dr.Bronner's, 1

WAX REMOVERS

General Retail Stores _____

Holloway House Wax Remover—Holloway House, Inc., 1

WOOD, PANELING, AND CABINET CLEANERS AND WAXES

General Retail Stores _____

Cabinet Magic—Magic American Corp., 1
Panel Magic—Magic American Corp., 1
Wood and Paneling Cleaner—Magic American Corp., 1

Home Shopping/Mail Order _____

Bekos-Bee & Resin Ointment—Livos Plant
    Chemistry, 1
Gleivo Liquid Wax—Livos Plant Chemistry, 1
Laro-Antique Wax—Livos Plant Chemistry, 1
Tekno-Cleaner—Livos Plant Chemistry, 1

APPENDIX 1
ORGANIZATIONS ACTIVE AGAINST ANIMAL PRODUCT TESTING

American Anti-Vivisection Society
801 Old York Road, Suite 204
Jenkintown, PA 19046-1685

Beauty Without Cruelty USA
175 W. 12th St., Suite 16G
New York, NY 10011

Friends of Animals (FOA)
P.O. Box 1244
Norwalk, CT 06856

The Humane Society of the United States (HSUS)
2100 L St., NW
Washington, D.C. 20037

The National Anti-Vivisection Society (NAVS)
57 West Jackson Blvd., Suite 1550
Chicago, IL 60604

New England Anti-Vivisection Society
333 Washington St.
Boston, MA 02135

People for the Ethical Treatment of Animals (PETA)
P.O. Box 42516 .
Washington, D.C. 20015-0516

APPENDIX 2
ALTERNATIVE FUNDING ORGANIZATIONS

These organizations actively search for and fund alternatives to the use of animals in product testing.

American Fund For Alternatives To Animal Research
175 West 12th St., Suite 16G
New York, NY 10011

International Foundation for Ethical Research (IFER)
53 Jackson Blvd.
Chicago, IL 60604

John Hopkins University Center for
  Alternatives to Animal Testing
615 N. Wolfe St.
Baltimore, MD 21205-2179

APPENDIX 3
TATTOO REGISTRATION
ORGANIZATIONS

The following organizations register tattooed pets. If your pet is lost or stolen and ends up in a research or testing lab, it can be returned to you through the tattoo and registration. It is best to place the tattoo on the inner thigh. Placing it in the ear will only encourage the removal of the ear itself. One-eared animals are a common sight in laboratories.

I.D. Pet
74 Hoyt St.
Darien, CT 06820
1-800-243-9147

National Dog Registry
P.O. Box 116
Woodstock, NY 12498-0116
1-800-637-3647 (1-800-NDR-DOGS)

North American Pet Owners Alliance, Inc.
P.O. Box 9202
Ft. Lauderdale, FL 33310
1-800-877-8729 (1-800-US-STRAY)

APPENDIX 4
RECOMMENDED READING

A Compendium of Alternatives to the Use of Live Animals in Research and Testing by Jeff Diner—Very comprehensive and in language easily understood; covers medical and drug testing as well as product testing (published by The American Anti-Vivisection Society, Jenkintown, PA; and the National Anti-Vivisection Society, Chicago, IL).

A Consumer's Dictionary of Cosmetic Ingredients by Ruth Winter—Lists the ingredients, toxicity, and— for some—even the source of the component; good reference for those concerned about product ingredients. (NY: Crown, 1976).

Alternatives to Current Uses of Animals in Research, Safety Testing, and Education by Martin Stephens, Ph.D.—Easily understood guide for the layman; covers all aspects of animal use and alternatives (Washington, D.C.: Humane Society of the United States, 1986).

Alternatives to Pain in Experiments on Animals by Dallas Pratt, M.D—Explains less painful ways of experimenting on animals as well as some alternatives to the use of animals; contains medical jargon and illustrates some truly horrifying uses of animals but is highly recommended (New York: Argus Archives, 1980).

Everything You Want to Know About Cosmetics by Toni Stabile. Although Ms. Stabile supports the use of animals in cosmetic testing, her book actually serves to show how useless these tests really are. This book is filled with stories of unsafe products that made it to the marketplace despite animal tests. Stories of manufacturers who ignored the results of their animal testing and released dangerous products are also found in this book (New York: Dodd, Mead, & Company, 1984).

The following magazines are also recommended to keep you up-to-date on all animal related issues:

Act'ion Line, published by Friends Of Animals, Norwalk, CT

The Animal's Agenda, published by the Animal Rights Network, Inc., Monroe, CT

HSUS News, published by The Humane Society of the United States, Washington, D.C.

NAVS Bulletin, published by the National Anti-Vivisection Society, Chicago, IL

PETA News, published by People for the Ethical Treatment of Animals, Washington, D.C.

APPENDIX 5
THE QUESTIONNAIRE

This is the questionnaire that was sent to the companies. (Changes were made in question 9 after the first 300 questionnaires were mailed, making the notarization voluntary instead of required.) The cover letter reiterated the need for ingredient lists and specifics on each product.

1. Does your company perform, or have performed for it, any consumer product testing that involves the use of animals?

2. Is this a permanent policy or a temporary moratorium?

3. Does this include all your products and product lines?

4. Does this include your parent or subsidiary companies?

5. Will this status change in the next six months?

6. Do your products contain ingredients that are derived from animals? If not, please include ingredient list.

7. Product categories will be further differentiated by where they are primarily sold. Please designate only one.

    a) Grocery/Drug/Department Stores
    b) Health Food Stores
    c) Home Shopping/Mail Order/Catalog Sales
    d) Other—such as private stores or hair salons. Please designate.

8. Is this the address you would like printed in the book? Would you like your phone number included also?

9. Please sign to verify your statements as true, notarize if possible. Include your position with the company.

10. Please list or attach complete product list. Be specific. Designate type of product if it cannot be surmised from name. Products will be grouped by use (shampoo, soap, floor cleaner, etc.) and type (liquid, bar, powder, etc.). If you are not specific I will have to guess.

APPENDIX 6
COMPANIES THAT DID NOT
RESPOND TO THE QUESTIONNAIRE

Abba Products
1800 Studebaker Road, Suite
585
Cerritos, CA 90701

Acclaim Entertainment
71 Audrey Ave.
Oyster Bay, NY 11771

Acu-trol, Inc.
2 Willow Road
St. Paul, MN 55127

Aditi-Nutri-Sentials
P.O. Box 155, Prince St.
Station
New York, NY 10012

Advance Design Laboratories
Box 55016, Metro Station
Los Angeles, CA 90055

African Bio-Botanicals, Inc.
7509 NW 13th Blvd.
Gainesville, FL 32606

Alfin Fragrances, Inc.
15 Maple St.
Norwood, NJ 07648

Almar Enterprises
P.O. Box 514
Farmington, MI 48332-0514

Almay Hypo-Allergenic
625 Madison Ave.
New York, NY 10022

Aloe Pro
2101 Midway Road #140
Carrollton, TX 75006

Aloe Up, Inc.
P.O. Box 2913
Harlingen, TX 78551

Alternative Cosmetics
P.O. Box 4052
Old Lyme, CT 06371

Alvin Last
145 Palisades St., Box 24
Dobbs Ferry, NY 10522

Alyssa Ashley, Inc.
1135 Pleasant View Terr. W
Ridgefield, NJ 53192

America's Finest Products
Corp.
1639 9th St.
Santa Monica, CA 90404

Amitol Industries/California
Cosmet
21100 Lassen St.
Chatsworth, CA 91311

Ananda Country Products
14618 Tyler Port Road
Nevada City, CA 95959

Andalina
Tory Hill
Warner, NH 03278-0057

Aspen Earth
1823 Main St.
Ferndale, WA 98248

Atta Lavi
443 Oakhurst Dr. #305
Beverly Hills, CA 90212

Austin Diversified Products
16615 S. Halsted St.
Harvey, IL 60426

Austin's
Box 827
Mars, PA 16046

Australian Gold
5811 W. Minnesota St.
Indianapolis, IN 46241

Ayurherbal Corp.
Box 2
Wilmot, WI 53192

Aztec Secret
Box 19735
Las Vegas, NV 89132

BeautiControl Cosmetics
600 Eagle Dr.
Bensenville, IL 60106

Beiersdorf, Inc.
360 Dr. Martin Luther King
    Dr.
P.O. Box 5529
Norwalk, CT 06856-5529

Bevan
P.O. Box 20072
New York, NY 10017-9993

Bo-Chem Co.
Little Harbor
Marblehead, MA 01945

Bonne Bell
Georgetown Row
Lakewood, OH 44107

Camilla Hepper/Baraka Co.
4338 Center Gate
San Antonio, TX 78217

Carme/Jojoba Farms/
    Loanda Soap/
    Biotene/Mill Creek
84 Galli Dr.
Novato, CA 94947

Caswell-Massey
111 Eighth Ave.
New York, NY 10011

C. E. Jamieson & Co., Ltd.
2 St. Clair Ave. W #1502
Toronto M4V1L5 Canada

Certan-Dri/Leon Products,
    Inc.
P.O. Box 4845
Jacksonville, FL 32241

Chae
815 E. 17th Ave.
Denver, CO 80218

Christian Dior Perfumes Inc.
9 W. 57th St.
New York, NY 10019

Citation Pet Products
930 SE. Lincoln
Portland, OR 97214

Clarins of Paris
540 Madison Ave.
New York, NY 10022

Clean Earth
2970 Blystone #102
Dallas, TX 75220

Color Quest
616 S. 3rd St.
St. Charles, IL 60174

Come To Your Senses
321 Cedar Ave. S
Minneapolis, MN 55454

Community Soap Factory
P.O. Box 32057
Washington, D.C. 20007

Cosmo Cosmetics
1101 S. Main
Los Angeles, CA 90015

Creative Nail Design
25255 Cabot Road #210
Laguna Hills, CA 92653

Creature Care
9009 S. Street, Box 763
Monte Rio, CA 95462

Critter Comfort
14200 Old Hanover Road
Reistertown, MD 21136

DeLore International
Box 37201
Cinncinnati, OH 45222

Dermatone Laboratories
47 Mountain Road, Box 633
Suffield, CT 06078

Details Intergrated
    Marketing
8282 Western Way Cir.,
    Unit A-4
Jacksonville, FL 32256

Diamond Products
435 Canning Road
Seffner, FL 33584

Dodge Chemical Co., Inc.
165 Rindge Ext.
Cambridge, MA 02140

Dont't Be Cruel
P.O. Box 46504
Chicago, IL 60646

Down To Earth Natural Foods
31 W. Main St.
Farmingdale, NJ 07727

Duncan Enterprises
5673 E. Shields Ave.
Fresno, CA 93727

Elizabeth Grady Face First
1 W. Foster St.
Melrose, MA 02176

Elvira Perfumes
9930 Pioneer Blvd, #101
Santa Fe Springs, CA 90670

Enhanced Water Products
8337 Penn Ave.
Minneapolis, MN 55431

Espree
P.O. Box 23448
Waco, TX 76702

Eve Cosmetics
P.O. Box 131
Pebble Beach, CA 93953

Fashion Two Twenty
1263 S. Chillicothe Road
Aurora, OH 44202

Faultless Starch/Bon Ami Co.
1025 W. 8th St.
Kansas City, MO 64101

Finnfoods
2355 Waukegan Road
Bannockburn, IL 60015

Fisher Price Toys
636 Gerard Ave.
E. Aurora, NY 14052

Flora Distributors
Box 67333
Vancouver, BC VSW 371
Canada

Fort Howard
1919 S. Broadway
P.O. Box 19130
Green Bay, WI 54307

Frank T. Ross and Sons, Ltd.
6550 Lawrence Ave.
P.O. Box 248
West Hill, ONT M1E4R5
Canada

Freeman Cosmetics Corp.
P.O. Box 4074
Beverly Hills, CA 90213

Friendly Systems
P.O. Box 1154
Euless, TX 76039

Fruit of the Earth, Inc.
P.O. Box 152044
Irving, TX 75015-2044

General Nutrition
921 Pennsylvania Ave.
Pittsburgh, PA 15222

Giorgio
9 W. 57th St.
New York, NY 10003

Giovanni Cosmetics, Inc.
P.O. Box 205
Reseda, CA 91335

Goldwell Cosmetics
9050 Junction Dr.
Annapolis Junction, MD
    20701

Greentree Grocers
3560 Mt. Acadia Dr.
San Diego, CA 92111

Guerlain
Rte. 138
Somers, NY 10589

H. Clay Glover Co., Inc.
1140 Franklin Ave.
Garden City, NY 11530

Hewitt Soap
333 Linden Ave.
Dayton, OH 45403

Houbigant
1135 Pleasant View Terr. W
Ridgefield, NJ 07657

House of Cheriss
P.O. Box 20359
Cleveland, OH 44120

Huish Chemical Co.
3540 W. 1987 S
Salt Lake City, UT 84125

I. Rokeach & Sons
25 E. Spring Valley Ave.
Maywood, NJ 07607

Ilona of Hungary, Inc.
3201 East 2nd Ave.
Denver, CO 80206

Image Laboratories
721 S. San Pedron
Los Angeles, CA 90014

Images & Attitudes
150 E. 400 N
Salem, UT 84653

Internatural Vitamin
Box 1746
Union, NJ 07083

James Austin
Box 827
Mars, PA 16046

Jean Naté
625 Madison Ave.
New York, NY 10022

Jean-Pierre Sand
1380 Queensgreen Cir.
Naperville, IL 60563

Jojoba Resources
6509 W. Frye Road #9
Chandler, AZ 85226

Joshua Solution
4151 N. 32nd St.
Phoenix, AZ 85018

Kleen Bright Laboratories
Box 20408
Rochester, NY 14602

La Coupe
694 Madison Ave.
New York, NY 10021

La Prairie, Inc.
600 Madison Ave.
New York, NY 10022

Lancaster
625 Madison Ave.
New York, NY 10022

Lange Products, Inc
21093 Forbes Ave.
Hayward, CA 94545

Larkspur
P.O. Box 40402
Pasadena, CA 91114

Life Essence
3438 E. Lake Road #14-655
Palm Harbor, FL 34685

Lilian Trading, Inc.
79 Belvedere St. #11
San Rafael, CA 94901

LJN Limited
1 Spring St.
Oyster Bay, NY 11771

Longevity Pure Medicine
9595 Wilshire Blvd. #706
Beverly Hills, CA 90212

Lowenkamp International
Monticello Road, Box 878
Hazlehurst, MS 39083

Luseaux Laboratories
16816 S. Gramercy Pl.
Gardena, CA 90247

Lydia O'Leary Covermark
1 Anderson Ave.
Moonachie, NJ 07074

M & N Natural Products
P.O. Box 4502
Anaheim, CA 92803

Make Up Art Cosmetics, Ltd.
233 Carlton St. #201
Toronto, Ontario M5A 2L2
Canada

Makimina, Inc.
P.O. Box 307
Wallingford, PA 19086

Marie Lacoste Enterprises,
Inc.
1059 Alameda de las Pulgas
Belmont, CA 94002

Martha Hill Cosmetics
5 Ivy Ct.
Metuchen, NJ 08840

Martin Von Meyering
422 Jay St.
Pittsburgh, PA 15212

Matrix Essentials
30601 Carter St.
Solon, OH 44139

Medical Plaza Consultants
4151 N. 32nd St.
Phoenix, AZ 85018

Mehron
45E Route 303
Valley Cottage, NY 10989

Merle Norman Cosmetics
9130 Bellanca Ave.
Los Angeles, CA 90045

Metrin Laboratories
1403-1275 Pacific St.
Vancover, B.C. Canada

Mill Creek
646-½ S. Venice Blvd.
Marina Del Rey, CA 90291

Mira Linder Spa In The City
29935 Northwestern Hwy.
Southfield, MI 48034

Naturade
7100 E. Jackson St.
Paramount, CA 90723

Natural Impressions
RD 2, Box 364A, Dewey Road
Cattaraugas, NY 14719

Natural Organics, Inc.
10 Daniel St.
Farmingdale, NY 11735

Natural Wonder
625 Madison Ave.
New York, NY 10022

Nature's Gate Herbal
    Cosmetics
9183-5 Kelvin St.
Chatsworth, CA 91311

New Age Creations/Herbal
    Bodyworks
219 Carl St. A
San Francisco, CA 94117

New Age Products
16100 N. Hwy. 101
Willits, CA 95490

New Leaf Market/Leon Co.
    Food Co-op
1235 Apalachee Pkwy.
Tallahassee, FL 32304

New World Minerals
4459 E. Rochelle Ave.
Las Vegas, NV 89121

Nexxus Products Co.
P.O. Box 1274
Santa Barbara, CA 93116

Nora's Natural Beauty
    Products

8039-D Penn Randall Pl.
Upper Marlboro, MD 20772

Nu Skin International, Inc.
145 East Center
Provo, UT 84606

Oleg Cassini/Perfums, Ltd.
3 W. 57th St., 8th Floor
New York, NY 10019

Panasonic
2 Panasonic Way
San Francisco, CA 07094

Park Rand Enterprises
12896 Bradley Ave. #F
Sylmar, CA 91342

People's Food Store
4765 Voltaire St.
San Diego, CA 92107

Phoenix Laboratories
    International
175 Lauman Lane
Hicksville, NY 11801

Physicians Formula
230 S. 9 Ave.
City of Industry, CA 91746

P. Leiner Nutritional
    Products
1845 W. 205th St., Box 2010
Torrance, CA 90510

Pro-Line
2121 Panoramic Cir.
Dallas, TX 75212

Purely Natural Body Care
Northrup Creek
Clatakanie, OR 97016

Queen Helene
100 Rose Ave.
Hempstead, NY 11550

Rainforest Fresh Organic
Herbs and Potpourri
36980 Wallace Creek Road
Springfield, OR 97477

Real Aloe Co.
P.O. Box 2770, 1620 Fiske Pl.
Oxnard, CA 93034

Redken Laboratories
6625 Variel Ave.
Canoga Park, CA 91303

RR Industries
1612 W. Olive, #301
Burbank, CA 91506

Rusk
1800 N. Highland #200
Los Angeles, CA 90028

Schiff
121 Moonachie Ave.
Moonachie, NJ 07074

Shirley Brown
17636 Corte Potosi
San Diego, CA 92128

Siri Skin Care
1719 N. Mariposa Ave.
Hollywood, CA 90027

Sixteenth Street Food Co-op
1318 Cortelyou Road
Brooklyn, NY 11226

Soap Works
60 Chatsworth Dr.
Toronto, Ontario M4R1R5
Canada

Solarex Consumer Products
Division
P.O. Box 6008
Rockville, MD 20850

SoRik International
278 Talleyrand Ave.
Jacksonville, FL 32202

St. Ives Laboratories, Inc.
8944 Mason Ave.
Chatsworth, CA 91311

Supreme Beauty Products
820 S. Michigan
Chicago, IL 60605

Teaco International Inc.

Tonialg Cosmetics
International
3095-E Presidential Dr.
Atlanta, GA 30340

Ultra Beauty
12233 S. Pulaski
Aslip, IL 60658

Universal Products
P.O. Box 580
Berryville, AR 72616

Val Chem
Box 338
Bayre, PA 18840

Vegetarian Concentrates
2105 Glendale Ave.
Sparks, NV 89431

Velvet Products Co.
P.O. Box 5459
Beverly Hills, CA 90210

Victoria Jackson
8500 Melrose Ave. #201
Los Angeles, CA 90069

Visage Beaute
9330 Civic Center Dr.
Beverly Hills, CA 90210

Vitamin Quota, Inc.
293 Madison Ave. #419
New York, NY 10017

Viviane Woodard Cosmetics
7712 Densmore Ave.
Van Nuys, CA 91406

Warner-Lambert Co.
201 Taboar Road
Morris Plains, NJ 07950

Watkins
150 Liberty St.
Winona, MN 55987-0570

Wella Corp.
525 Grand Ave.
Englewood, NJ 07631

West Cabot Cosmetics Assoc.
165 Oval Dr.
Central Islip, NY 11722

White King
P.O. Box 2198
Terminal Annex
Los Angeles, CA 90051

Youthessence, Ltd.
P.O. Box 3057
New York, NY 10185

Yves Rocher
Yves Rocher Center 2672
West Chester, PA 19380-2672

APPENDIX 7
CATALOG COMPANIES THAT DID NOT RESPOND TO THE QUESTIONNAIRE

A Clear Alternative
8707 West Lane
Magnolia, TX 77355

Ayagutag
Box 176
Ben Lomond, CA 95005

Co-Op America
2100 M St. #403
Washington, D.C. 20063
202-872-5307

Compassion Cosmetics
Box 3534
Glendale, CA 91201

Humane Street U.S.A.
467 Saratoga Ave., Suite 300
San Jose, CA 95129
408-243-2430

Kind Kare, Inc.
134 W. University, Suite 125
Rochester, MI 48063

My Brothers Keeper
P.O. Box 1769
Richmond, IN 47375
317-962-5079

Nature Basics
61 Main St.
Lancaster, NH 03584
603-788-4432

Red Saffron
3009 16th Ave. S
Minneapolis, MN 55407
612-724-3686

NOTES

1. U.S. Department of Health and Human Services, Food and Drug Administration, *Requirements of Laws and Regulations Enforced by the U.S. Food and Drug Administration* (Washington, D.C.: Government Printing Office, 1989), p. 56.
2. "Tracking 'Buyers Beware'" *NAVS Bulletin* 1 (1988), p. 10.
3. Toni Stabile, *Everything You Want to Know About Cosmetics*, (New York: Dodd, Mead, 1984), pp. 38-41.
4. *The PETA Guide to Animals and Product Testing*, (Washington, D.C.: People for the Ethical Treatment of Animals), p. 5
5. Stabile, *Everything You Want To Know*, p. 56.
6. Dallas Pratt, M.D., *Alternatives to Pain in Experiments on Animals*, (New York: Argus Archives, 1980), p. 206.
7. Pratt, *Alternatives to Pain*, p. 174
8. Paul Penders, "Cruelty-Free Cosmetics," *Feather River Catalog 1990*, p. 197.
9. Penders, "Cruelty-Free Cosmetics," p. 196.
10. "Cosmair Still Testing," *PETA News* (September/October 1989), p. 6.
11. Ibid., p. 203.
12. Mike McIntire, "Lost And Found: The Tragic Fate of Stolen Pets," *Act'ion Line* (April/May 1990), p. 7.
13. Ibid., p. 6.
14. ABC News, "20/20" (July 13, 1990).
15. "Tragic Fate," p. 6.
16. "Tragic Fate," p. 8.
17. "Law Notes, 'The HSUS, ALDF Sue USDA,'" *HSUS News* (Fall 1990), p. 36.
18. Ibid., p. 36.
19. Ibid., p. 36.
20. News Shorts, *The Animal's Agenda* (September 1990), p. 43.
21. Jeff Diner, *A Compendium of Alternatives to the Use of Live Animals in Research And Testing* (Jenkintown, PA and Chicago, IL: The American Anti-Vivesection Society and The National Anti-Vivisection Society), p. 14.

22. Martin L. Stephens, Ph.D., *Alternatives to Current Uses of Animals in Research, Safety Testing, and Education* (Washington, D.C.: The Humane Society of the United States, 1986), p. 17.
23. "Tracking, 'Important Reminder,'" *NAVS Bulletin 1* (1988), p. 24.
24. *PETA Guide*, p. 6.
25. "LD50 Tests," *PETA Factsheet #9* (Washington, D.C.: People for the Ethical Treatment of Animals).
26. Pratt, *Alternatives to Pain*, p. 206.
27. Donald J. Barnes, "Circadian Rhythms Invalidate Research," *NAVS Bulletin* 1 (1988), p. 5.
28. Pratt, *Alternatives to Pain*, p. 207.
29. "Classical LD/50," *HSUS Fact Sheet* (Washington, D.C.: The Humane Society of the United States, 1984).
30. Ibid.
31. "LD50 Tests."
32. *PETA Fact Sheet #10* (Washington, D.C.: People for the Ethical Treatment of Animals).
33. Pratt, *Alternatives to Pain*, p. 219.
34. Ibid., p. 220.
35. *Requirements of Laws and Regulations*, p. 56.
36. "Compassion Campaign," *PETA Guide to Compassionate Living* (Washington, D.C.: People for the Ethical Treatment of Animals), p. 7.
37. Diner, *Compendium of Alternatives*, p. 9.
38. Lisa Lockard, "Putting an End to a Dying Tradition," *NAVS Bulletin* 1 (1990), p. 16.
39. *The Use of Animals in Product Safety Testing* (Washington, D.C.: Foundation for Biomedical Research, 1988), p. 7.

About the Author

Lori Cook has devoted her life to improving animal welfare. She has a degree as a veterinary medical technician and spent many years working for a veterinary firm, during which time she became familiar with the workings of not only the veterinary industry, but horse and dog racing, zoos, animal pounds, shelters, laboratories, and farms. She supports numerous animal and environmental organizations and does volunteer work at the Humane Society. She is a vegetarian, does not wear leather or fur, and tries to shop cruelty-free. She lives in Tampa, Florida with her two mutt dogs.

If you know of any companies we've missed, please write to:

5364 Ehrlich Road
Box 501
Tampa, FL 33625